LEEDS BECKETT UNIVERSITY
LIBRARY
70 0040797 4

ROOF CONSTRUCTION FOR DWELLINGS

ROOF CONSTRUCTION FOR DWELLINGS

FIRST EDITION REVISED

C.N. Mindham, BSc

Foreword by Jack A. Baird, CEng, MIStructE

BSP PROFESSIONAL BOOKS

OXFORD LONDON EDINBURGH

BOSTON MELBOURNE

First Edition published by Collins
Professional and Technical Books 1986
First Edition Revised published by
BSP Professional Books 1988
Reprinted 1989

British Library
Cataloguing in Publication Data

Mindham, C. N. (Chris N.)
Roof construction for dwellings. – Rev.
1. Roofs. Design & construction
I. Title
695

ISBN 0–632–02308–2

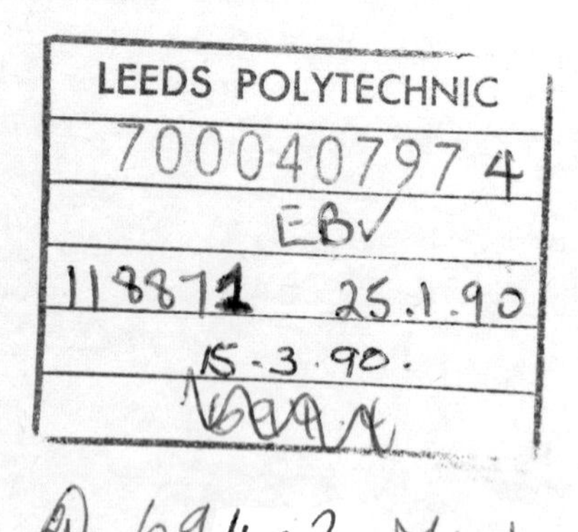

BSP Professional Books
A division of Blackwell Scientific
Publications Ltd
Editorial Offices:
Osney Mead, Oxford OX2 0EL
(Orders: Tel. 0865 240201)
8 John Street, London WC1N 2ES
23 Ainslie Place, Edinburgh EH3 6AJ
3 Cambridge Center, Suite 208, Cambridge,
MA 02142, USA
107 Barry Street, Carlton, Victoria 3053,
Australia

Set by Columns of Reading
Printed and bound in Great Britain by
Richard Clay Ltd, Bungay, Suffolk

Contents

Foreword

My early involvement with structural engineering was mainly with steelwork with which, whether for power stations or refugee dwellings, it is usual for all details, including the size and length of bolts, to be worked out before components are delivered to site. Consequently, when I became involved with timber engineering and timber components for small buildings with masonry walls, I was surprised by the general lack of detailed information on connections provided by the building or specialist designer, and the reliance placed on site operatives and those useful inventions – the saw, nail and hammer – to invent connections on site.

In fairness, designers are often encouraged by the client to spend as little time as possible in producing drawings, which can be why detail is not covered. However, sometimes the lack of detailed drawings is linked to a lack of detailed knowledge. After all, how many designers in timber can provide off-the-cuff details of the most efficient way of forming a hip end or valley condition, etc., in a trussed rafter or traditional roof?

Chris Mindham has had plenty of experience of such construction and, therefore, I am delighted that he has taken time to detail solutions which he has found to work in practice. As more of our future designers and decision makers learn their engineering from theory rather than practical drawing-board situations, such information will become even more valuable. In the meantime, I see this book as being useful shelf reference for anyone in their first forty years or so in the house building game.

Jack A. Baird
CEng, MIStructE

Preface

My experience over many years working with architects, engineers, surveyors, building managers and site staff indicated a need for a book covering construction practice for roofs and in particular those of domestic housing. Like many other building topics the roof is one of those subjects with which everyone is familiar until it comes to actually detailing or cutting the timber components concerned, and then the lack of knowledge becomes apparent. Furthermore, research soon confirmed the total lack of in-depth text on the construction of trussed rafter roofs, a method of construction now used on 90% of house construction in the United Kingdom.

The book aims to describe with the aid of many drawings, not the structural design analysis of the roof structure, but the design of the roof assembly as a whole entity rather than individual elements in isolation. Recognising the growing trend to refurbish older homes, the traditional or 'cut' roof is described. The bolted and connectored roof is dealt with in some detail, for despite the popularity of the trussed rafter this older system is still chosen by some builders. The bolt and connector truss roof is particularly popular for small extension projects where it often continues the construction of the original roof.

Chapters 5 and 6 cover the trussed rafter roof in great detail, dealing with the often misunderstood hip construction, valleys, girder truss assemblies, and the forming of openings in roofs as well as attic constructions. Chapter 6 compares the various truss plate systems and has been made as accurate as possible, bearing in mind the many changes being introduced by these manufacturers to their engineering services and computer programs with the current introduction of British Standard 5268 part 3.

It is the intention that the book be used for reference, and to this end there is a small degree of repetition between chapters, and there is frequent cross-referencing between chapters for both text and illustrations. Although some basic common knowledge of building is anticipated, most terms used are fully described, making the book equally suitable for use by both the building student and the professional. The text takes into account the latest issues of both the British Standard for timber engineering British Standard 5268 parts 2 and 3 and the Building Regulations 1985. However, as it was felt to be outside the scope of this book, the subject of fire resistance and spread of flame has not been dealt with. Reference should be made to Building Regulation Approved Documents.

For ease of reference all drawings have been given a number, the first digit of which refers to the chapter, and the second and third digits being the numerical sequence in that chapter. Generally, shading has been used to highlight those elements discussed in the text to which the illustration applies. Most drawings have been produced in perspective form to aid quick appreciation of the three-dimensional nature of all roof structures. Chapter 2 sets out the terms used throughout the book to describe roof and truss shapes, and individual roof members. The specialised terminology of the trussed rafter is given in Fig. 5.2. Finally for those involved in the design aspects of roof structures, the British Standard 5268 parts 2 and 3 should be available for ready reference.

This revised edition takes into account amendment No 1 to British Standard 5268 part 3, published in 1988.

C. N. Mindham
August 1988

Acknowledgements

I would like to thank those people who, however fleetingly, have helped me in research and preparation of this book. Special thanks are due to TRADA and in particular Maurice Sawyer for his assistance with Chapter 4 and encouragement to write the book. Thanks are also due to Gang-Nail Limited, in particular to John Gordon for his assistance with Chapter 5, and to the Building Department and the library of the West Suffolk College of Further Education for their free assistance at research stage.

Chapter 6 has been produced with the willing co-operation of Messrs Bevplate Limited, Gang-Nail Limited, Hydro-Air Limited, Trusswal Limited and Twinaplate Limited, all of whom have given freely of their technical information and to whom I am indebted for the use of their various illustrations in this chapter.

Finally my thanks are due to my wife for tolerating the sometimes not inconsiderable mess of paper, literature and drawings cluttering the family home.

CHAPTER 1
The development of the pitched roof

PRIMITIVE ROOF FORMS

Man has always needed a roof for shelter. Early man used roofs formed by nature such as caves, but nomadic peoples had to be more resourceful, creating shelters of a temporary nature each time they moved. It is likely that simple tents formed with animal skins over branches were the early form of constructed roofs, with more permanent shelter being pit dwellings. These were simply a shallow excavation covered with a simple roof of branches and skins. It is an easy step from this type of dwelling to a simple wall on the edge of the pit to raise the headroom and then to use shaped branches to give a slight pitch, thus improving rain run-off and therefore the quality of the environment within the shelter.

The simple 'cruck' frame comprised two curved pieces of timber standing on the ground at one end and meeting at the top. Across several of these 'crucks' were tied horizontal members onto which, again, were fixed skins or as time progressed simple thatch.

THE COUPLED ROOF

Moving away from early roof forms that provided both wall and roof in one unit, the next development showed a true roof built on masonry or timber walls. The simplest form of roof was a coupled roof, consisting of two lengths of timber bearing against each other at the top and resting on a wall plate at their feet. The timbers, called couples, were pegged together at the top with timber dowels and were similarly pegged or spiked to the wall plate. The term 'couple' was used until the fifteenth century when the terms 'spar' or 'rafter' started to be used. The term rafter of course is still used to describe the piece of timber in a roof spanning from the ridge to the wall plate.

The couples were generally spaced about 400 mm apart tied only by horizontal binders and tile battens. The simple couple was adequate for small span dwellings and

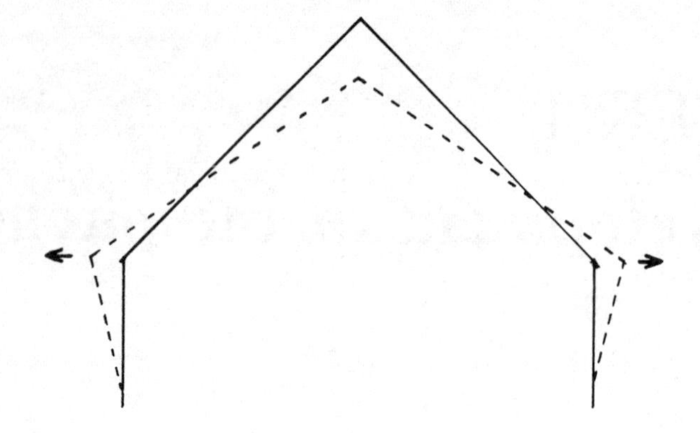

Fig. 1.1

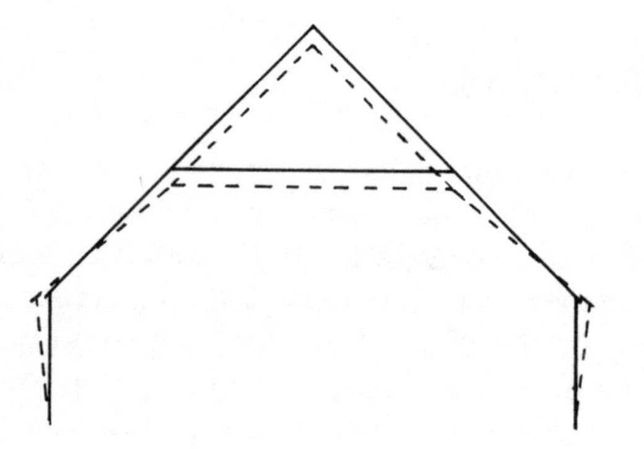

Fig. 1.2

steep pitches, but the outward thrusting force at the feet of the rafters caused stability problems with the walls, and excessively long rafters sagged in the middle under the weight of the roof covering. The illustration in fig. 1.1 shows the required shape in solid line and the deflected shape in dotted line.

To overcome both of these problems the 'wind beam' or 'collar' was introduced. Whether the collar acts as a tie or a strut for the couples will depend upon the stiffness of the supporting wall below. Assuming however, that the wall is so substantial that it will not be pushed outward by the bottom section of the couples, then the collar will act as a strut. If however, as is more likely with early timber framed buildings, the walls would be relatively flexible then in that case the collar would act as a tie holding the couples together. There would still be some outward thrust but this would be limited by the collar to the degree of bending in the lower part of the couple only. Fig. 1.2 illustrates this condition. It can readily be appreciated that in larger roofs, where the walls are relatively flexible, there is a considerable tying effect in the collar demanding a more sophisticated joint between collar and couple than could be achieved with simple iron nails. The collar was therefore frequently jointed to the couple with a halved dovetail shaped joint, often secured with hardwood pegs.

STABILITY

The next development was to fit additional members to assist with the stability of the roof in windy conditions and these were called 'sous-laces' or braces. On roofs constructed on substantial masonry walls which were also very thick, further struts or 'ashlars' were introduced to stiffen the lower section of the couple. Fig. 1.3 illustrates this form of construction, the wall plate being well fixed to the wall with the bottom member of the ashlar halved over it to prevent the roof sliding on the top of the wall.

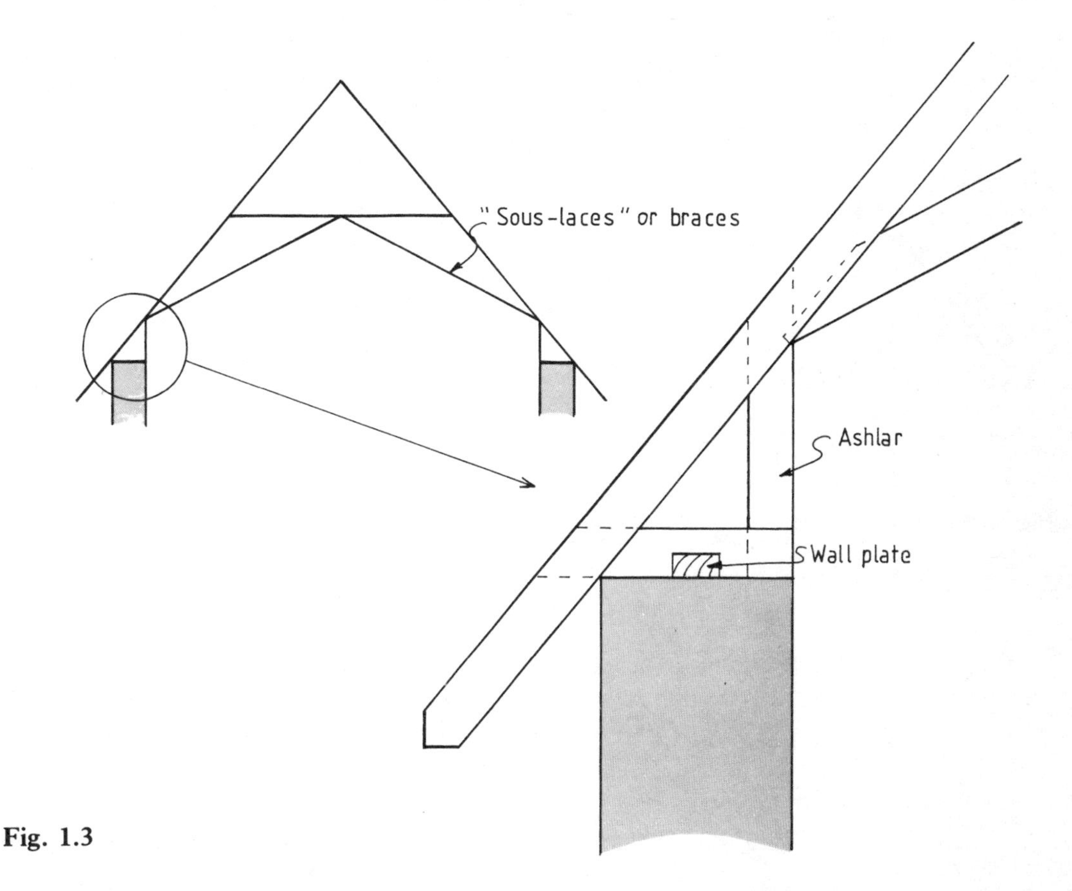

Fig. 1.3

These now very substantial 'couples' began to be spaced further apart and became known as 'principals'. Between these main members simple couples or 'rafters' were placed, but to avoid sag or to accommodate longer rafter length possibly not available in one length of timber, an intermediate support was needed and this was called a 'purlin'. The purlin is in turn supported by the principal couples, as shown in fig. 1.4.

The tendency for the roof to spread was now concentrated in the heavily loaded principals and it became apparent that if spans were to increase this spreading would have to be controlled. The 'tie beam' was introduced thus forming the first 'trussed' or 'tied' roof. Fig. 1.5 illustrates the roof form described.

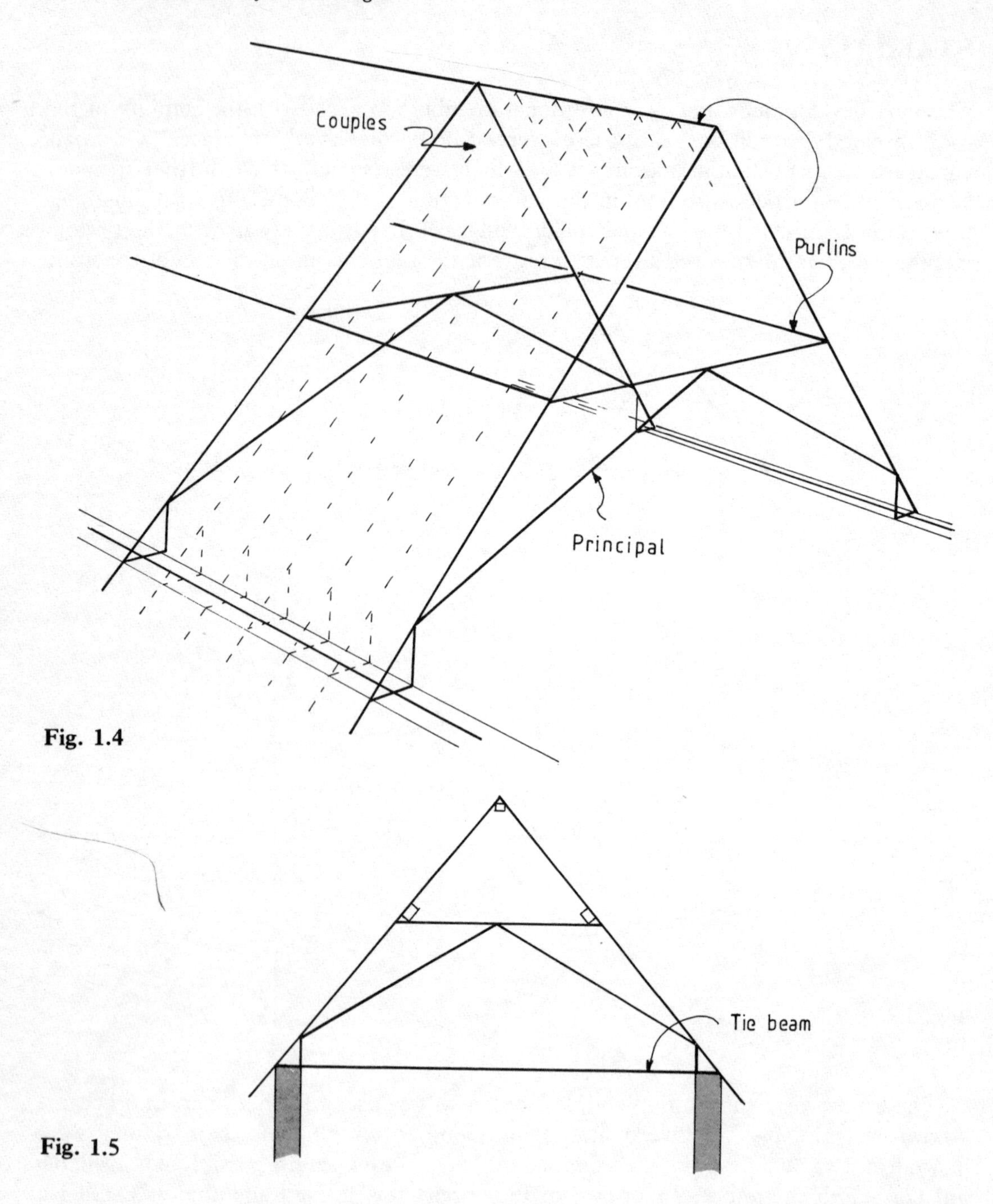

Fig. 1.4

Fig. 1.5

As development progressed the span of the roof was limited only to the availability of long timbers used for the tie beam, but it is obvious that these long beams themselves would tend to sag under their own weight. To prevent this happening they too had to be supported and this was done with the introduction of 'struts' fitted to a corbel built into the wall below, as illustrated in fig. 1.6.

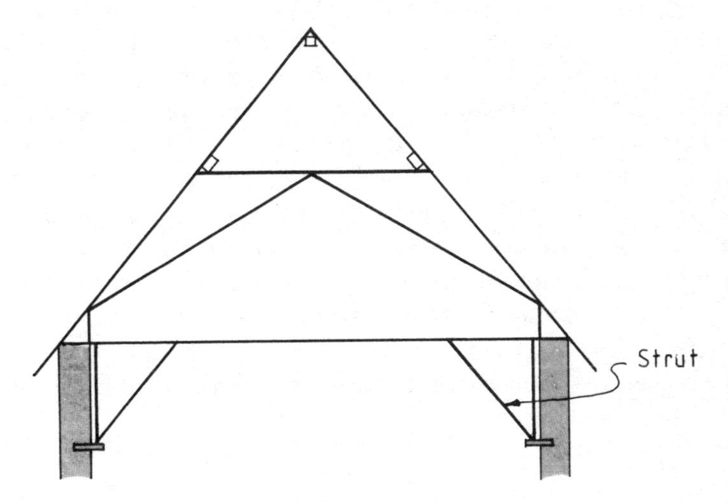

Fig. 1.6

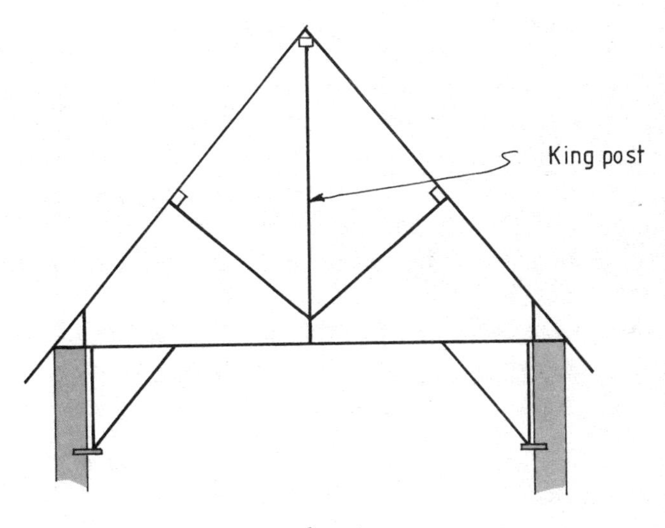

Fig. 1.7

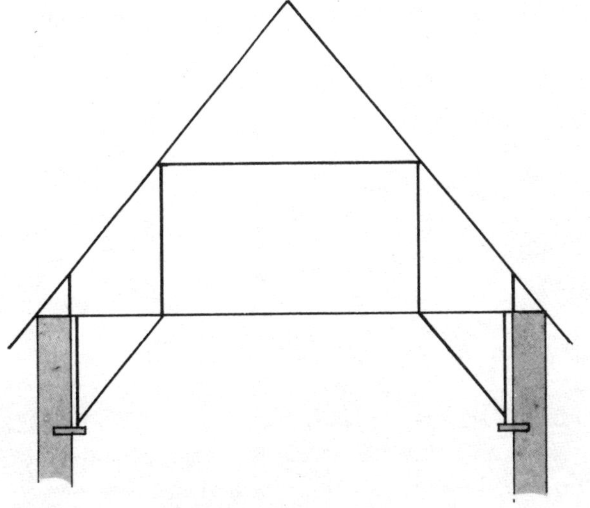

Fig. 1.8

With this tie beam now becoming a major structural member a different configuration of members evolved becoming more like the truss common today. Having stiffened the tie beam it became apparent that this could be used as a major structural item from which to support the principals. The major support running from the centre of the tie beam to the ridge purlin was known as the 'mountant' now referred to as a 'king post' (see fig. 1.7). With two posts introduced the roof form is known as a 'queen post' truss, which in its simplest form is shown in fig. 1.8. This particular roof form gave the opportunity of providing a limited living space within the roof. It should be remembered that until this stage of development all roof forms and trusses described had no ceiling and were open to the underside of the rafters and roof covering. To use the queen post roof form as an attic, a floor was needed thus creating a ceiling for the room below.

CEILINGS

Ceilings were first referred to in descriptions of roofs in the fifteenth century when they were known as 'bastardroofes' or 'false roofs' and then later as 'ceiled roofs', hence 'ceiling' as we know it today.

The ceiling supports were known as joists or cross beams again being supported by the hard working tie beam between the principals.

Continuing developments of the roof form itself, and demand for even larger spans and heavier load resulted in some relatively complex principals or trusses being developed. One such form was the 'hammer beam' roof, illustrated in fig. 1.9. Clearly this is not a roof to be 'ceiled', being very ornate as well as functional.

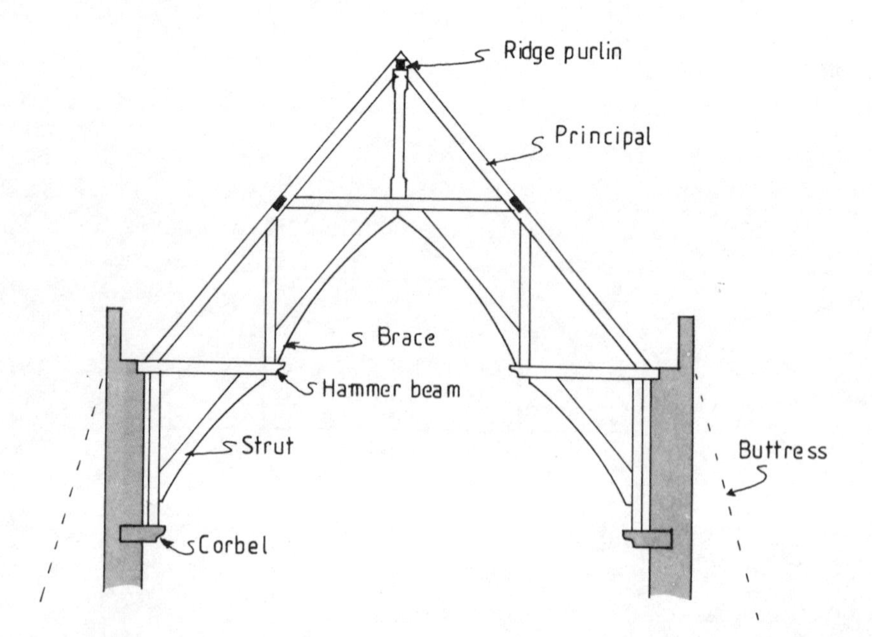

Fig. 1.9

The hammer beam roof is generally to be found supporting the roof over halls in large mansions and of course in churches. The roof was framed in such a way as to reduce the lateral thrust without the need for a large and visually obstructing tie beam. The walls onto which such a roof was placed had to be substantial and were often provided with buttresses in line with the principals to contain any lateral thrust that may develop.

TRUSSES

Roofs in truss form developed using carpentry joints and some steel strapping, until the latter part of the eighteenth century when bolts, and even glues, started to be used to create large truss forms from lighter timber members. Such truss forms often used softwoods, as distinct from the hardwoods more frequently used in the shapes previously described. The large timber sections in oak particularly were becoming very scarce and of course very expensive. Whilst some significant advances in span were achieved, using the techniques described above, the domestic roof did not require very large spans and changed very little from the collared coupled roof. Indeed many small terraced houses built during the eighteenth and nineteenth century required no principals at all. The dividing walls between the houses were close enough to allow the purlins to rest on these walls, effectively using them as principals. Fig. 1.10 illustrates a typical terraced house roof construction.

The larger properties where the span of the purlin was too long for one piece of timber, or where hip ends were involved, continued either to use the established methods of construction using principals, collars, and purlins, but it was common practice to omit the principals and to support the purlins off the walls below with posts or struts.

DESIGN FOR ECONOMY

In 1934 the Timber Development Association (TDA) was formed, now known as TRADA (Timber Research and Development Association). The Association took up the work already being done at that time by the Royal Aircraft Establishment and progressed work on timber technology alongside the Forest Product Research Laboratories. Although the Royal Aircraft Establishment may sound a strange body to be interested in timber, it must be remembered that many aircraft of that era, and some notable ones after such as the Mosquito, used highly stressed timber structures for the fuselage and wings. Some aircraft hangars were of timber construction and utilised record breaking large span small timber section trusses with bolted joints.

After the Second World War shortages of materials resulted in a licence being required for all new building works, making economy in use of paramount importance. Imported materials such as timber were very much at a premium and TDA was given the task to find ways of economising on the country's use of timber. Quite correctly

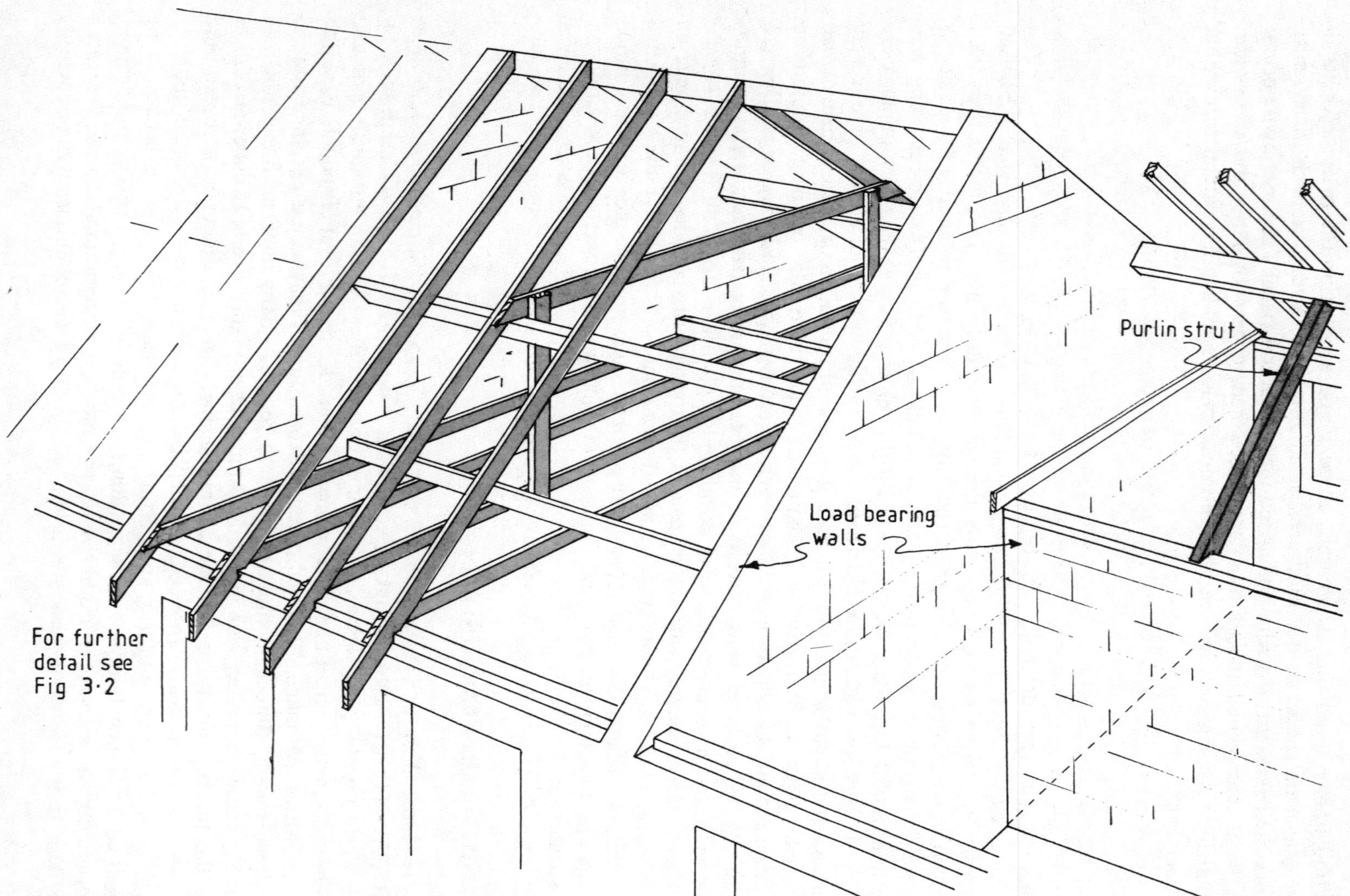
Purlin strut
Load bearing walls
For further detail see Fig 3·2

Fig. 1.10

they identified the roof structures of buildings as a high volume user of timber and developed a design for a domestic roof using principal trusses constructed of small timber sections connected with bolts and metal connector plates. The roof used purlins and common rafters similar to the systems previously discussed. These trusses became known as 'TDA' trusses, and with some minor modifications are still in use today. It appears that some of these designs were available shortly after the Second World War but were first published as a set of standard design sheets around 1950.

The designs were based on existing truss shapes but were not engineered in the sense that structural calculations were prepared for each design. Load testing on full size examples of the truss was used to prove their adequacy and from these tests other designs developed.

STANDARD DESIGN ROOFS

The first designs produced were known as 'A' and 'B' types, dealing with 40° and 35° pitches respectively. They covered spans up to 30 ft (9 m).

House design fashion changed during the later fifties and early sixties, demanding lower roof pitches. 1960 saw the introduction of the TDA type 'C' range for pitches between 22° and 30°. Spans were also increased up to 32 ft (10.8 m). Around 1965 the types 'D', 'E' and 'F' ranges were published, these later designs using a slightly different truss member layout went down to 15° pitch and up to 40 ft (12 m) span. Further designs used trusses spaced at 6 ft (1.8 m) centres and had some degree of pitch and span flexibility within specified limitations.

A range of designs for trussed rafters (i.e. each couple tied together at ceiling level) was produced also using bolt and connector joints, but these were designed only to carry felt roof coverings and did not prove as popular as the principal truss designs. Industrial roofs were not neglected, with principal truss designs using the bolt and connector joint techniques for pitches of 22.5° spacing between 11 and 14 ft (3.35-4.25 m) and up to 66 ft (20.1 m) span.

BOLT AND CONNECTOR JOINTS

All of the TDA principal and trussed rafter designs used bolts and connectors at joints where previously mortice and tenon, half lap or straight nailed or pegged joints would have been used. The small timber sections used in the designs of the trusses did not allow the use of conventional carpentry joints and gave insufficient nailing area for an all nailed assembly. The connector allows the forces in the joint to be spread over a large area of the connected timber, the bolt holding the timbers in place thus allowing the connector to transmit the load from one truss member to the other. Fig. 1.11 illustrates the typical single connector joint.

All of the designs mentioned above are available today, and these are complemented by span tables for the design of purlins. TRADA is therefore a valuable source of

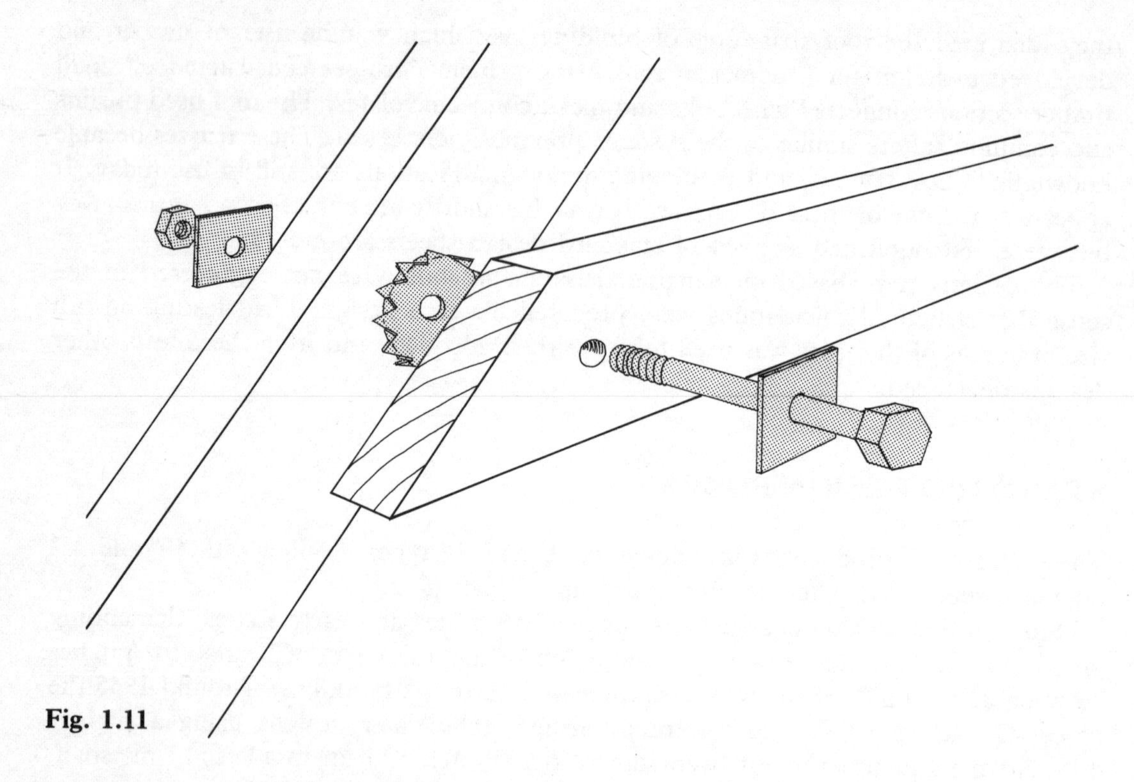

Fig. 1.11

information for those constructing timber roofs. A more recent design, currently being developed by TRADA is a 'stressed skin' panel design for 'room-in-the-roof' construction. This uses common rafters formed into panels by fixing them together with thin plywood covering. The panels then lean together at the ridge and are fitted to a special shaped plate fixed to the timber floor and are stiffened by a collar bolted into position. This particular design of course lends itself to factory made panels and fast and efficient on-site construction.

TRUSSED RAFTERS

In the early 1960s the punched metal connector plate was introduced into the UK from the USA and was to revolutionise the construction of domestic roofs even more than the TDA truss designs described. There are now five main plate manufacturers in the UK, the first in 1967 being Gang-Nail whose name has come to be used to describe all punched metal connector trusses, in the same way that 'Hoover' seems to describe a vacuum cleaner.

Trussed rafters are generally prefabricated in a factory and transported to site, although with certain types of plate, fabrication can take place on site. In the case of metal plates, the manufacturer sells plates backed up to varying degrees with design aids to approved manufacturers, many of whom are also timber merchants. The timber used is both graded for strength and machined on all surfaces to give accuracy to the finished product. Trussed rafters can also be assembled using plywood gussets, the plywood being either nailed to a defined pattern or nailed and glued to the truss members to form the joint. Standard designs for such trussed rafters are available from the Swedish Finnish Timber Council. Ply gussetted trusses are not as popular as metal plated trusses, but do offer a method of manufacture not requiring specialist equipment. Similarly the galvanised steel plates punched with a pattern of holes to receive nails can also be used to form truss joints and these too can be fabricated on site.

Fig. 1.12

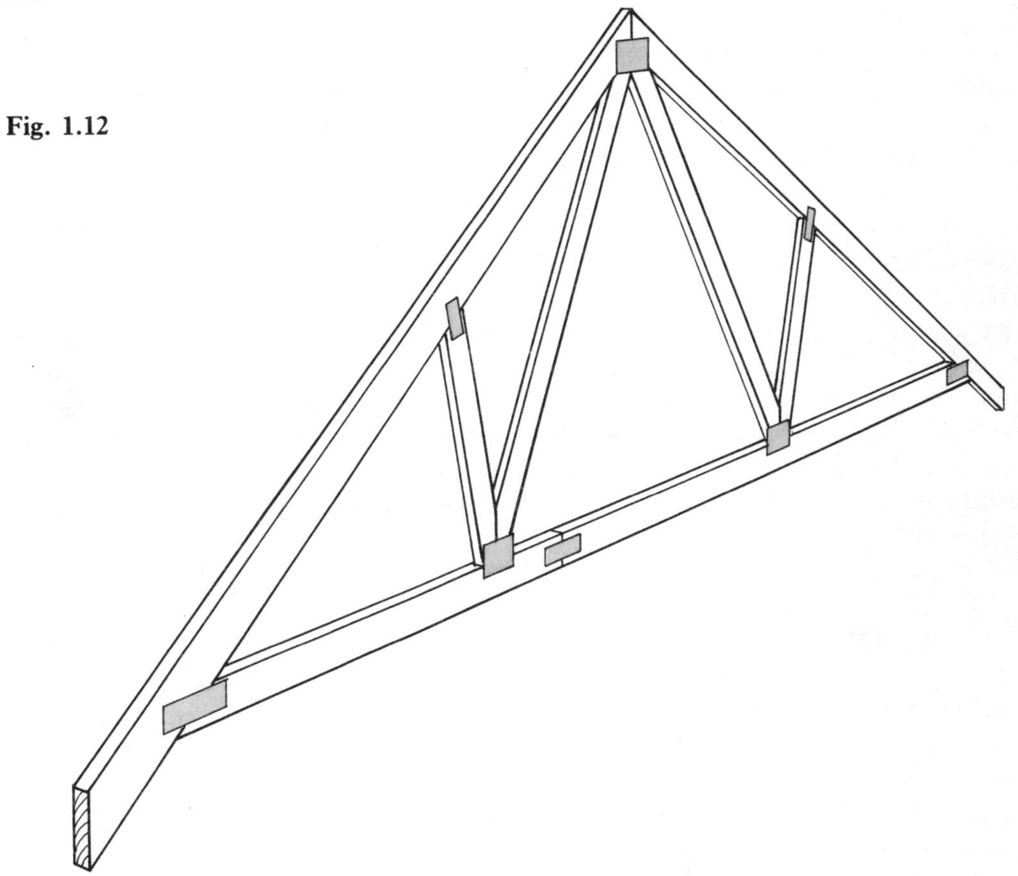

The punched metal nail plates used in factory trussed rafter production are mechanically pressed into the timbers on both sides of each joint to form a trussed rafter. This trussed rafter is then placed on the roof at approximately 600 mm centres taking the place of the common rafter. Hence its term 'trussed rafter', as distinct from the TRADA type principal truss, although it will be seen later in chapter 5 that trussed rafters themselves can be used to form principal or girder trusses. A typical 'fink' trussed rafter is illustrated in fig. 1.12.

COST ADVANTAGES

Trussed rafters are designed to carry simply the direct load imposed upon them. It is assumed that they are to be kept upright by other members, these members being the binders and diagonal bracing and even the tile battens vital to the overall stability of the roof. Whilst most trussed rafters are used for roofs of housing, their use is increasing for roofs of public buildings, commercial buildings and to a lesser extent for industrial and agricultural buildings. Clear spans in excess of 30 m can be achieved with lightweight roof coverings.

When first introduced into the UK, the designs were limited to those contained in standard design manuals, thus the duo pitch and mono pitched roofs were common but more complex roofs needed individual designs prepared. The advent of the computer both speeded up and dramatically reduced the cost of the design process, and this has been further advanced by the use of microcomputers installed in many trussed rafter manufacturers' offices. There are now almost no limitations to the possible shape of trussed rafters, except those imposed by the practicality of production and transportation to site.

Trussed rafter roofs use approximately thirty per cent less timber than a traditional roof, and can be built into a roof form in a fraction of the time taken for either a truly traditional common and purlin roof, or a TRADA construction. Factory production keeps the labour cost of trussed rafter manufacture very low compared to that necessary to assemble a bolt and connector jointed truss, thus giving further cost advantages to the trussed rafter. Almost all new housing now uses a trussed rafter form of roof construction.

LEGISLATION

Having looked at the development of the roof form, we must take account of the legislation controlling building construction in the UK. Before this century no controls existed, and it was not until the introduction of the model byelaws by each local authority area that some degree of control was placed upon the design of buildings.

The Building Regulations as we now know them first appeared in 1965, and have been amended and re-issued on several occasions since that date. Subsequent amendments have dealt with such roof related matters as the restraint of gable end walls, thermal insulation and roof void ventilation. Now re-issued as the Building Regulations 1985, we have an informative and fairly easily understood document compared to the earlier issues of the regulations. This latest issue came into force on 11 November 1985. These regulations lay down the legal requirements for building and concern themselves with health and safety aspects and not the aesthetic aspects of the structure. The latter, of course, is controlled by the local planning authorities.

The National House-Building Council (known as NHBC) has its own set of standards, which although incorporating the Building Regulations requirements, look beyond the health and safety aspects and seek to set minimum standards for quality control and such items as heating, electrical power sockets, and the general finish given to the buildings. Formed in 1936 it was not until the mid-sixties that the council began to have influence on the vast majority of house builders in the UK.

Concerned by the so-called 'gerry builders' after the Second World War, the building societies needed some method of ensuring that the homes on which they had granted mortgages were of an adequate standard to protect their investment. These societies therefore demanded that house builders building and wishing to sell homes on which the societies were granting mortgages must belong to the NHBC and submit themselves to their inspections. Having achieved full compliance with the NHBC requirements and of course the Building Regulations, the mortgage would be granted. Consequently most newly built homes until now have had to be inspected by the local authority as well as the NHBC, although this is likely to change in the near future, and only the inspectorate of the NHBC will be involved. An alternative to NHBC for mortgage purposes in most instances, is that the house should be inspected by a registered architect, and this seems to be the only way that a non-registered house builder can build and sell a new home under a mortgage agreement.

The Building Regulations and NHBC standards in turn refer to various *British Standards* and it is intended here only to deal with those British Standards concerned with timber in roof structures.

Code of Practice number 112 started life in 1952, and was amended in 1967 when the principle of allocating grade stresses to timber was introduced. 1971 saw further changes to the code then issued with stresses and timber sizes in metric units. In 1973 Code of Practice number 112 part three 'Trussed Rafters for Roofs of Dwellings' was first published, and is now re-issued as British Standard number 5268 part 3 1985. Code of Practice 3: Chapter V: Part 1: 1967 *Loading. Dead and imposed loads* (soon to be issued as BS 6399: Part 1 in rewritten form) is also relevant and CP3: Chapter V: Part 2: 1972 *Loading. Wind loads*, is mandatory. BS 5250 concerning roof void ventilation and BS 5534: Part 1 on slating and tiling must also be taken into consideration.

Sawn softwood timber sizes are set out in British Standard 4471 giving the standard range of timber sizes and tolerances therefrom, whilst British Standard 4978 'Timber Grades for Structural Use' issued in 1983 deals with stresses allocated to various timbers for structural design purposes. The latter document is recommended for those wishing to have some insight into the type of timber that they can expect within the various stress gradings. Such factors as knots, fissures, bow, spring and twist are all dealt with, giving limiting factors; pages 11-18 will be found to be of great benefit to students.

All structural timber used in dwellings must now be graded into stress limiting classes and marked with the grades. This mark must show not only the grade, but the grader and grading station number, the British Standard number and the species group.

Grading can be carried out either visually by qualified visual graders, or by a licensed stress grading machine operated by trained staff. To aid the appreciation of grading in ascending order of strength for any one species we have a list set out below

- GS: General structural
- MGS: General structural (machine graded)
- M 50: Timber with defects limiting its strength to approximately 50% of a defect-free piece of timber of the same section
- SS: Select structural
- MSS: Select structural (machine graded)
- M 75: Timber in which the defects are considered to limit it to no less than 75% of the strength of a defect-free piece of timber of the same section.

These classifications are given to each species or group of species, and it can therefore be seen that there are many permutations of species and grades available, leading unfortunately to poor understanding of the real potential of timber as a structural material.

The latest British Standard BS 5268 part 2 1984 has simplified this relatively complex subject by grouping timbers into strength classes. These range from strength class SC1 for the weakest timbers to strength class SC9 for those extremely strong hardwoods. Most timbers used for roof structures will fall in the SC3 and SC4 strength classes and indeed it is these two strength classes which form the basis of the tables in the latest issue of the building regulations.

As can be seen from the short discussion above, timber grading is a detailed scientific subject well outside the scope of this book. The book is concerned with the constructional design of the roof, rather than the calculation of structural design and specification. The reader is directed to those British Standards referred to above, the many publications available from TRADA and the Swedish Finnish Timber Council, and the *Timber Designers' Manual* by Baird and Ozelton.

Roof development will undoubtedly continue. The timber sizes used in modern trussed roof construction really constitute the practical minimum possible. Structurally it may be feasible to reduce those sizes, but for reasons of achieving adequate fixings for ceiling boards and tile battens, the timber sizes cannot be reduced. It is therefore difficult to see beyond the trussed rafter, but its method of construction into the completed roof form may change.

Although the labour involved in erecting a trussed rafter roof is relatively small, access at roof plate level and within the roof structure whilst under construction is not good. As will be seen in Chapter 5, the new British Standard requires a considerable amount of additional bracing to be installed within the roof, thus increasing the labour involved and the hazards of gaining access within the roof void. For this reason a practice relatively common with the large panel timber framed housing systems may become increasingly popular. This is to construct the roof including wall plate at ground level, complete with all binders, bracing, ties, tank platform, tank, felt battens, barge and fascias where appropriate. This whole, relatively light assembly can then be craned on to the shell and fixed in position. It is not suggested that this is a cost effective method for very small building sites, but on the larger estates, where continuity of house building is achieved, it has many advantages, not the least of which is the safety of the workman concerned.

CHAPTER 2
Roof shapes and terminology

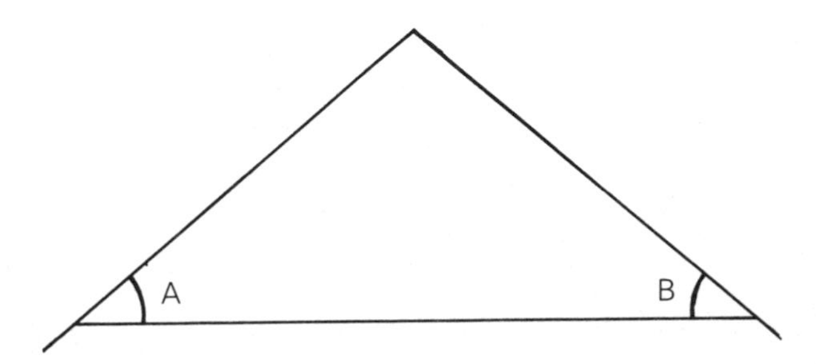

Fig. 2.1 Duo-pitched roof. This is the most common roof shape with equal pitches on either side, i.e. angle A equals angle B.

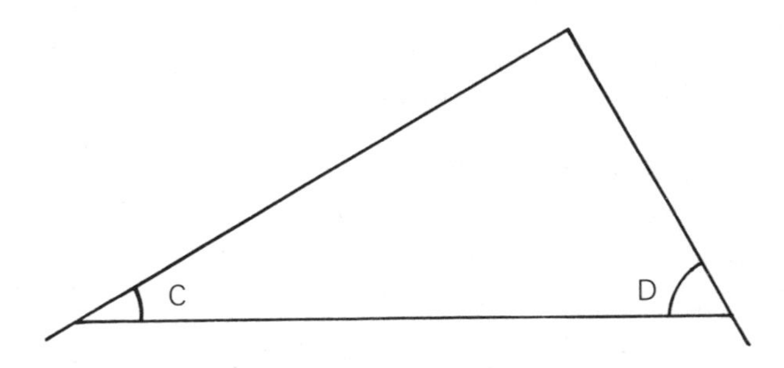

Fig. 2.2 Asymmetric roof; angle C is not equal to angle D.

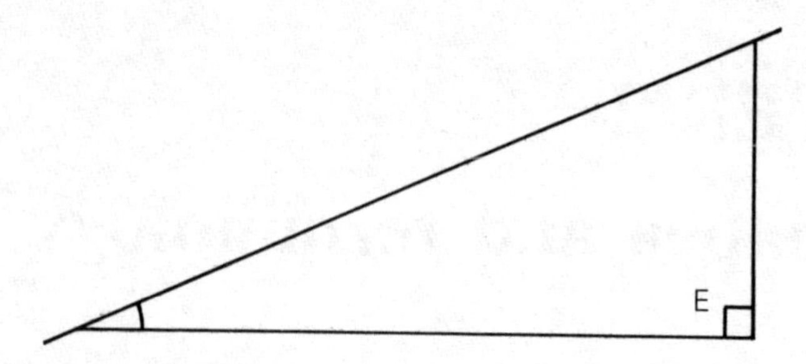

Fig. 2.3 Mono pitched roof; angle E equals 90°.

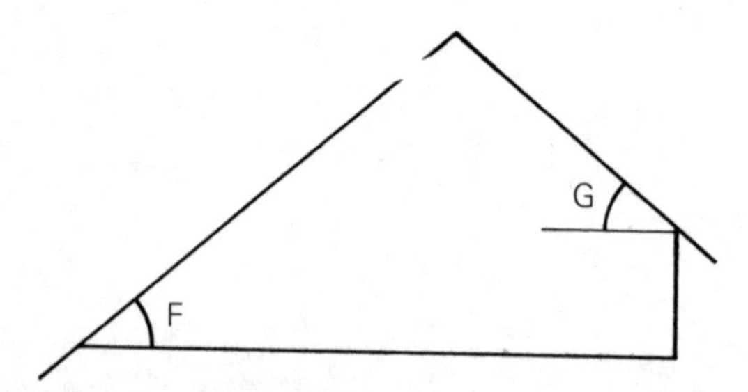

Fig. 2.4 Truncated duo pitched roof; angle F equals angle G. This truss form is often introduced into domestic housing in conjunction with the conventional duo pitched roof to form an interesting roof line.

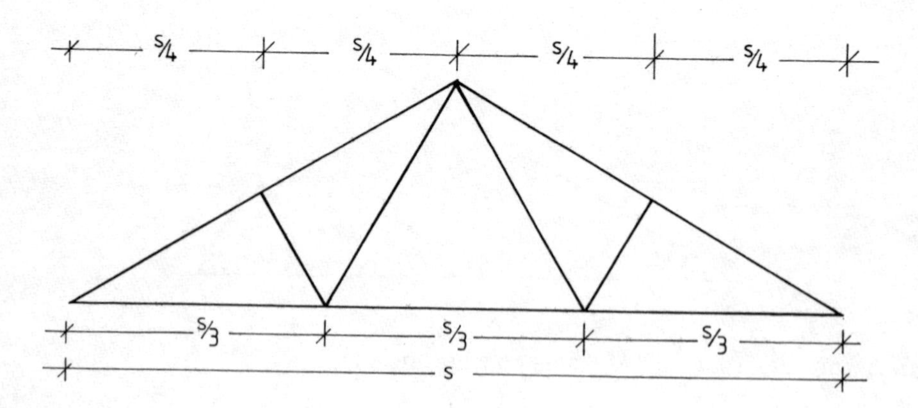

Fig. 2.5 Fink truss shape. This is the most common trussed rafter form used on spans of up t 8 to 9 m.

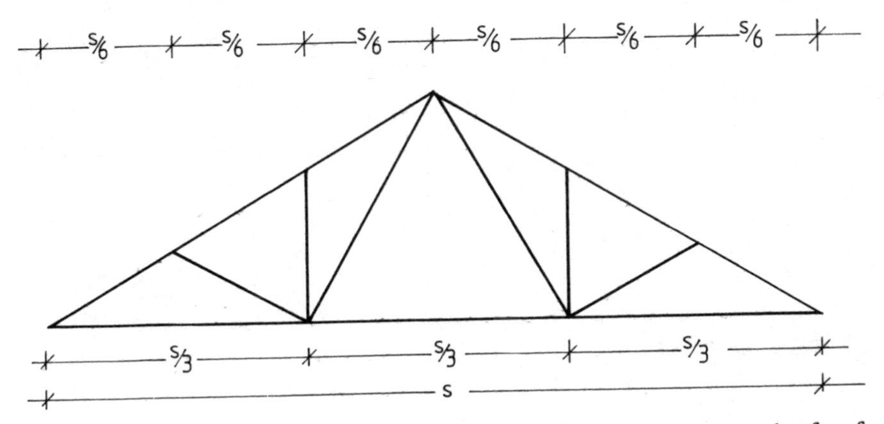

Fig. 2.6 Fan truss shape. This is used on larger spans and is a common trussed rafter form.

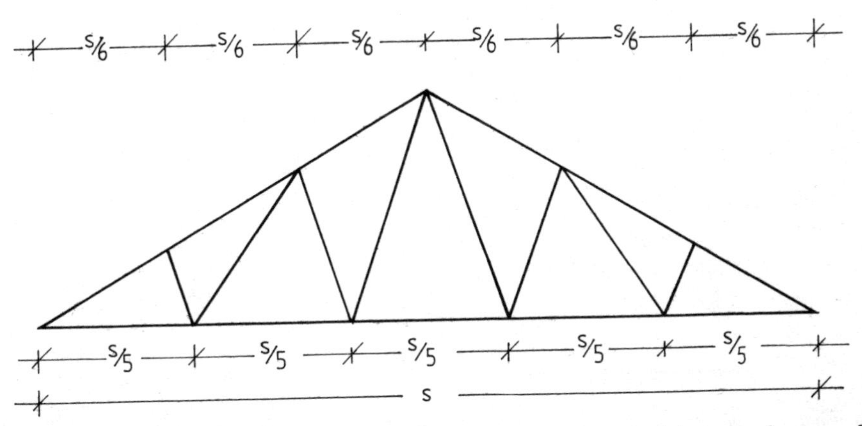

Fig. 2.7 Double 'W' shape. This is used on spans above 14 m and is not often used on housing.

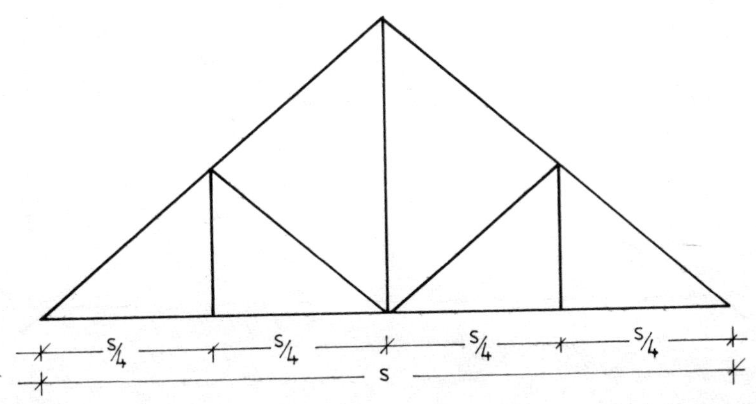

Fig. 2.8 Howe four bay truss. This is often used in trussed rafters in girder form. This could also be used in six bay configuration.

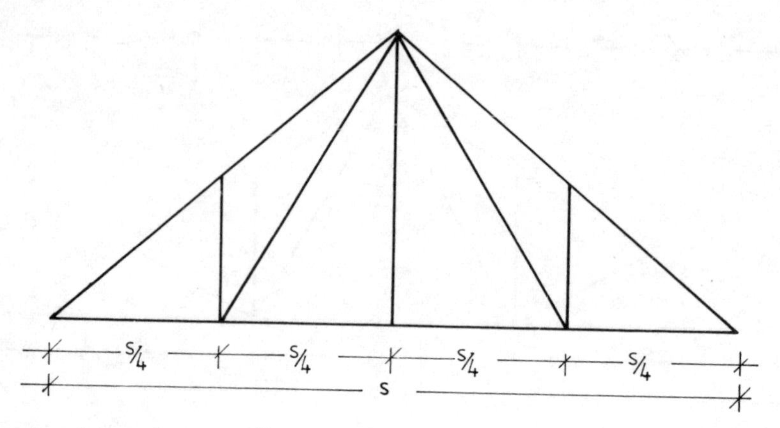

Fig. 2.9 Pratt four bay truss. This is occasionally used in trussed rafters in girder form.

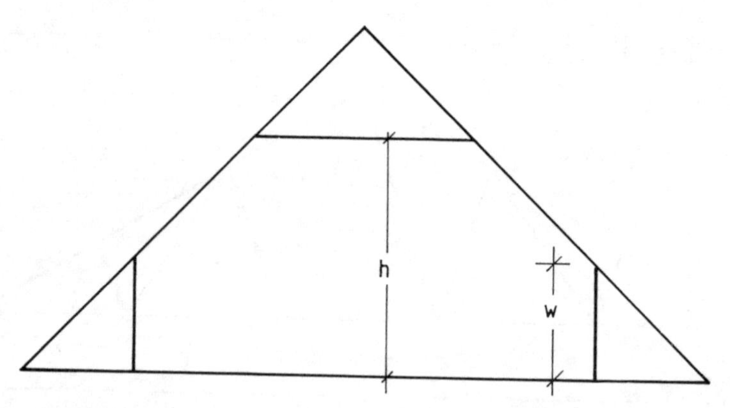

Fig. 2.10 Attic or 'room-in-the-roof' truss shape. This is a popular shape in trussed rafters; h: minimum 2.3 m, W: no minimum height set, but from 1.2 to 1.5 m is a practical minimum.

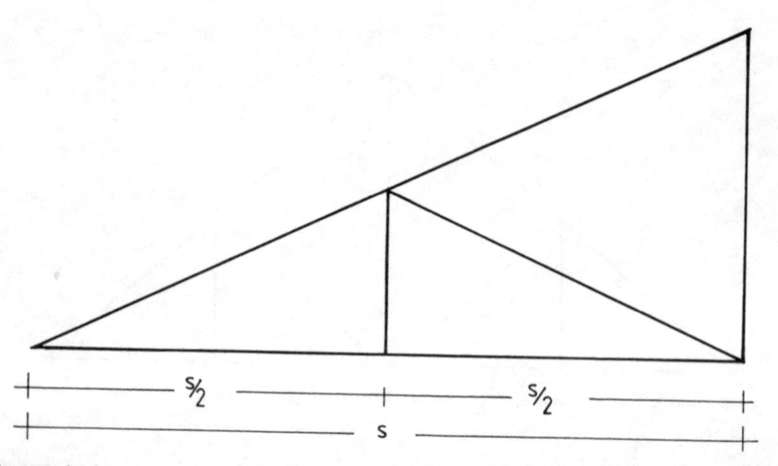

Fig. 2.11 Mono pitch truss two bay. This is a common trussed rafter form often used in conjunction with trusses in figs 2.5 to 2.7.

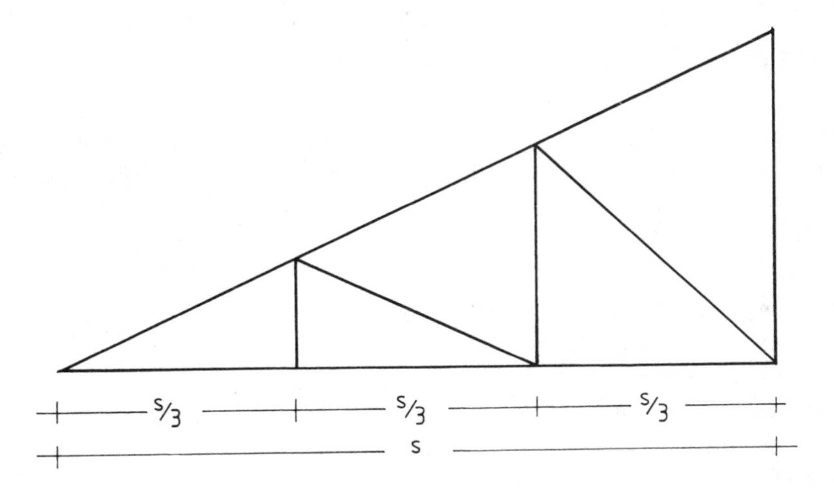

Fig. 2.12 Mono pitch truss three bay. This is similar to the mono pitch truss two bay (fig. 2.11), but is suitable for larger spans.

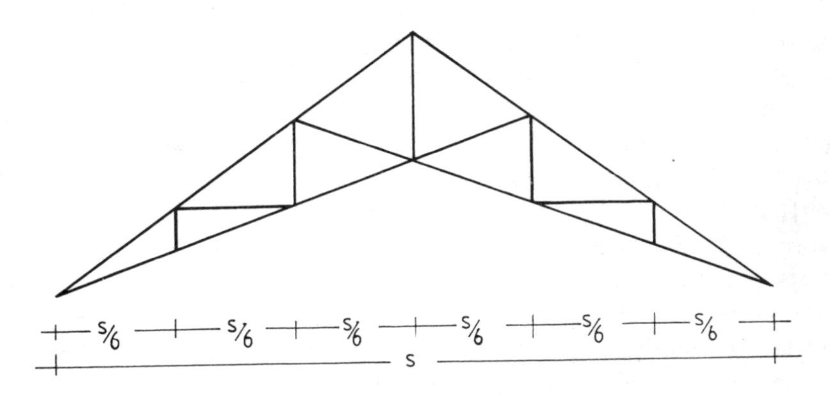

Fig. 2.13 Scissor truss. This is a possible trussed rafter shape occasionally used to create a feature ceiling in the lounge of a house.

TERMINOLOGY

Chapter one has given the history and derivation of some of the names given to roof structure members. The list below, although not exhaustive, describes the terms used on domestic roof structures.

The reader is referred to fig. 2.14.

A – Wall plate – sawn timber, usually 50 × 100 or 50 × 75 mm bedded in mortar on top of the inner skin of a cavity wall. Straps must be used to secure the wall plate to the structure below (see Chapter 7, figs 7.8 and 7.9).

B – Common rafter – sawn timber placed from wall plate to ridge to carry the loads from tiles, snow and wind. Long rafters may need intermediate supports from purlins.

B1 – Jack rafters – sawn timber rafter cut between either a hip or valley rafter (see Chapter 3, fig. 3.7).

C – Ceiling joist – sawn timber connecting the feet of the common rafter at plate level. The ceiling joist can also be slightly raised above the level of the wall plate, but this would technically then be termed a collar. The ceiling joist supports the weight of the ceiling finish (normally plasterboard) and insulation. It may in addition have to carry loft walkways and water storage tanks, in which case it must be specifically designed to do so.

D – Ridge – a term used to describe the uppermost part of the roof. The term is also used to describe the sawn timber member which connects the upper parts of the common rafters.

E – Fascia – usually a planed timber member used to close off the ends of the rafters, to support the soffit M, to support the last row of tiles at the eaves N and to carry the rainwater gutter support brackets.

F – Hip end – whereas a gable end O is a vertical closing of the roof, the hip is inclined at an angle usually to match the main roof.

F1 – Hip rafter – sawn timber member at the external intersection of the roof slope (similar to a roof sloping ridge), used to support the jack rafters forming the hip (see Chapter 3, fig. 3.7).

G – Valley – term used to describe the intersection of two roofs creating a ‘valley’ on either side. The illustration has only one main valley, the building being L-shaped on plan. A further small valley is illustrated on the dormer roof with its junction to the main roof. Valley jack rafters are fitted either side of a valley rafter, as illustrated in Chapter 3, fig. 3.10.

H – Dormer – the structure used to form a vertical window within a roof slope (see Chapter 3, fig. 3.17 for other shapes of dormer). This structure gives increased floor area of full ceiling height within an attic roof construction, and is usually fitted with a window, hence the term ‘dormer window’.

I – Barge board – this piece of planed timber is in fact a sloping fascia. It is often fitted to gable ends, as illustrated.

J – Dormer cheek – the term used to describe the triangular infill wall area between

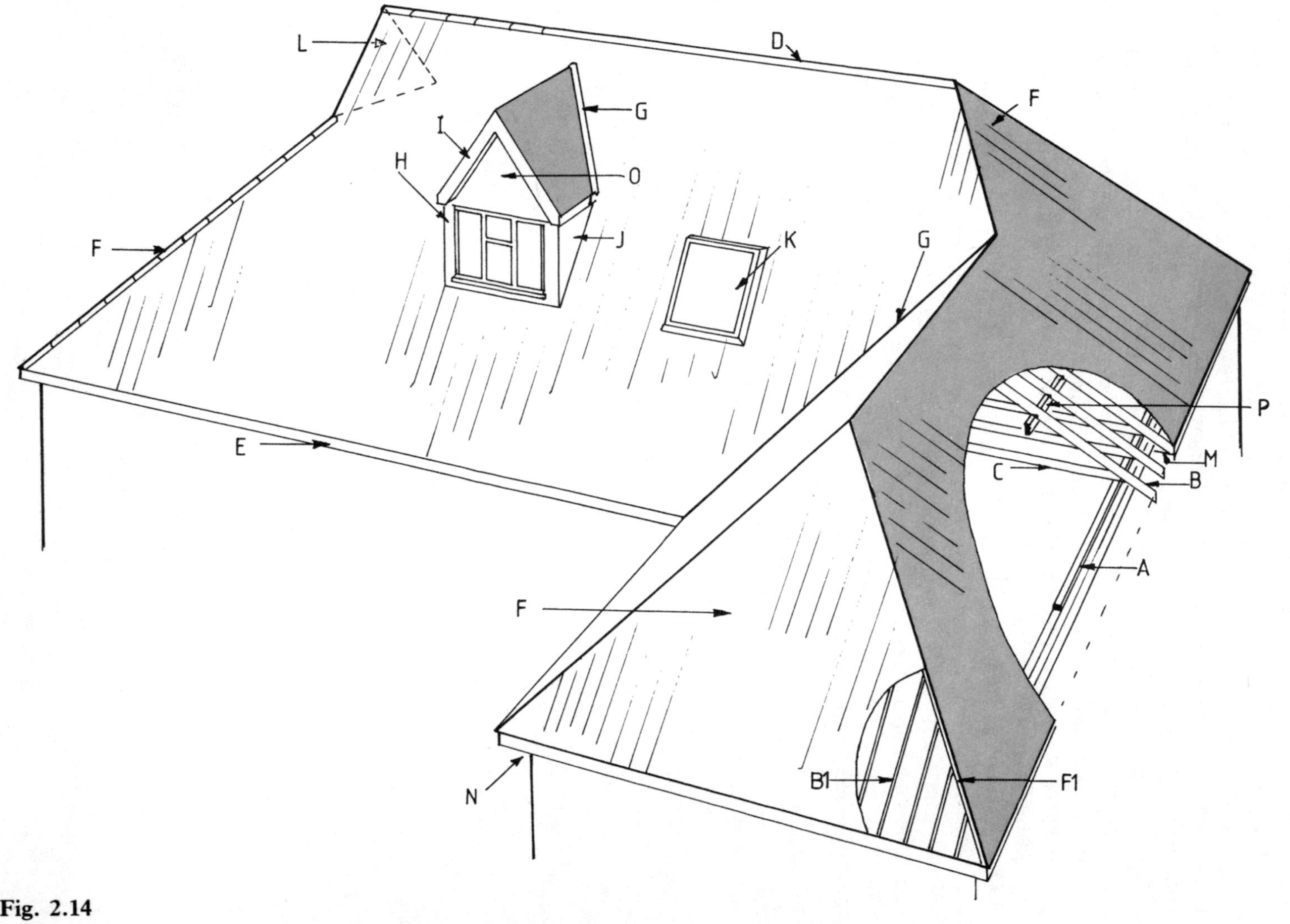

Fig. 2.14

dormer roof, main roof and the dormer front (see Chapter 3, fig. 3.18 for the construction).

K – Roof window – sometimes termed roof light, the former being able to be opened for ventilation hence becoming a true window, the latter being fixed simply allowing additional light into the attic roof space.

L – Gablet – a small gable over a hip end. It is used as an architectural feature.

M – Soffit – the ply or other sheet material panel used to close off the space between the back of the fascia and the wall of the building.

N – Eaves – term used to describe the extreme lower end of the roof, i.e. the area around the fascia and soffit.

O – Gable – triangular area of wall used at the end of a roof to close off beneath the roof slopes. This is usually a continuation of the wall construction below.

P – Purlin – large section sawn solid structural timber, or fabricated beam, used to carry the common rafters on larger roof slopes where the commons are not strong enough or cannot be obtained in one single length, to span between the wall plate and the ridge (see figs 3.2, 3.5 and 3.6).

CHAPTER 3
The 'traditional' or 'cut' roof

DESIGN

The traditional or 'cut' roof as it has become known is essentially a roof cut and assembled on site from individual timber members. It is most frequently a common rafter and purlin roof, the design of which can be prepared from readily available standard span tables for the individual timber members. Hips and valleys are generally constructed to what has become known as 'good practice' and are less well documented with span tables and specific design aids. The sizing of these members is often left to the architect or engineer and it is not always necessary to provide calculations to prove their adequacy. The design of all new roof structures in England, Wales and Inner London must of course conform with the requirements of the Building Regulations 1985. In Scotland the existing Building Standards (Scotland) Regulations 1981 to 1984, will continue to be used, although some amendments will be introduced during 1986. Northern Ireland will continue to use the Building Regulations (Northern Ireland) 1977 (as amended).

Whilst the Building Regulation statutory document provides the functional requirements with which one has to conform, the approved documents contain many span tables which, if used to design the individual timber members, will ensure compliance with the functional requirements and will be structurally sound. A guide to the timber members to be found in these tables and the limiting pitches is given in fig. 3.1.

Other design aids can be obtained from TRADA, the Council of Forest Industries, the Canadian Timber Promotion Organisation, known as COFI and the Swedish Finnish Timber Council. Based on this information a roof may be designed and built for most common roof shapes.

THE COMMON RAFTER AND PURLIN ROOF

This simple form of roof is illustrated in fig. 3.2. The structure is most commonly used where there is a gable both ends of the roof, and is frequently to be found on terraced

Fig. 3.1

	ROOF PITCH		
Roof Member	10° – 22½°	22½° – 30°	30° plus – 42½°
Purlin	B10/B12	B14/B16	B18/B20
Rafter	B9/B11	B13/B15	B17/B19
Ceiling Joist	B5/B7	B5/B7	B5/B7
Ceiling Joist Binder	B6/B8	B6/B8	B6/B8
Floor Joist	B3/B4	B3/B4	B3/B4
Full Data In Building Regulations 1985 Approved Document A Part B Tables B3 – B28			

houses, as indicated in fig. 1.10. Its construction has been included here because of the now very common refurbishment of such houses.

The wall plates are often simply bedded on mortar on either the inner skin of a cavity wall or, as is often the case with older terraced houses, on the inside edge of a solid 9 in brick wall. Wall plates should be half lapped where they meet, should not be less than 75 mm wide and 50 mm thick. They should be treated with preservative. Fig. 3.3 shows typical plate connections. Further reference should be made to Chapter 7 where wall plates are dealt with in detail.

Purlins

In some of the older houses purlins were placed at right angles to the rafter. A more effective construction results with the purlins truly vertical for three reasons:

(1) The purlin is easier built-in or set in hangers at the gable walls.
(2) The purlin allows the rafter to be bird's mouthed over them, thus avoiding the tendency for the rafter to slide off the roof. A notch in the rafter can be used on sloping purlins but a birdsmouth is easier to locate and a quicker joint to cut on site.
(3) The sloping purlin has a tendency to sag down the roof slope thus necessitating a much thicker timber to maintain a true line. Fig. 3.4 illustrates this point.

A common problem with this type of roof is the tendency to stretch the purlins structurally close to their design limit, so achieving maximum economy on the section of the purlin to be used. This sometimes can result in roof sag caused by deflection of the purlin, although the deflection may be within design tolerances. There are two ways of overcoming this problem. One is to design a stiffer purlin, i.e. probably one or two sizes up from the design table solution, the other is to stiffen the purlin using

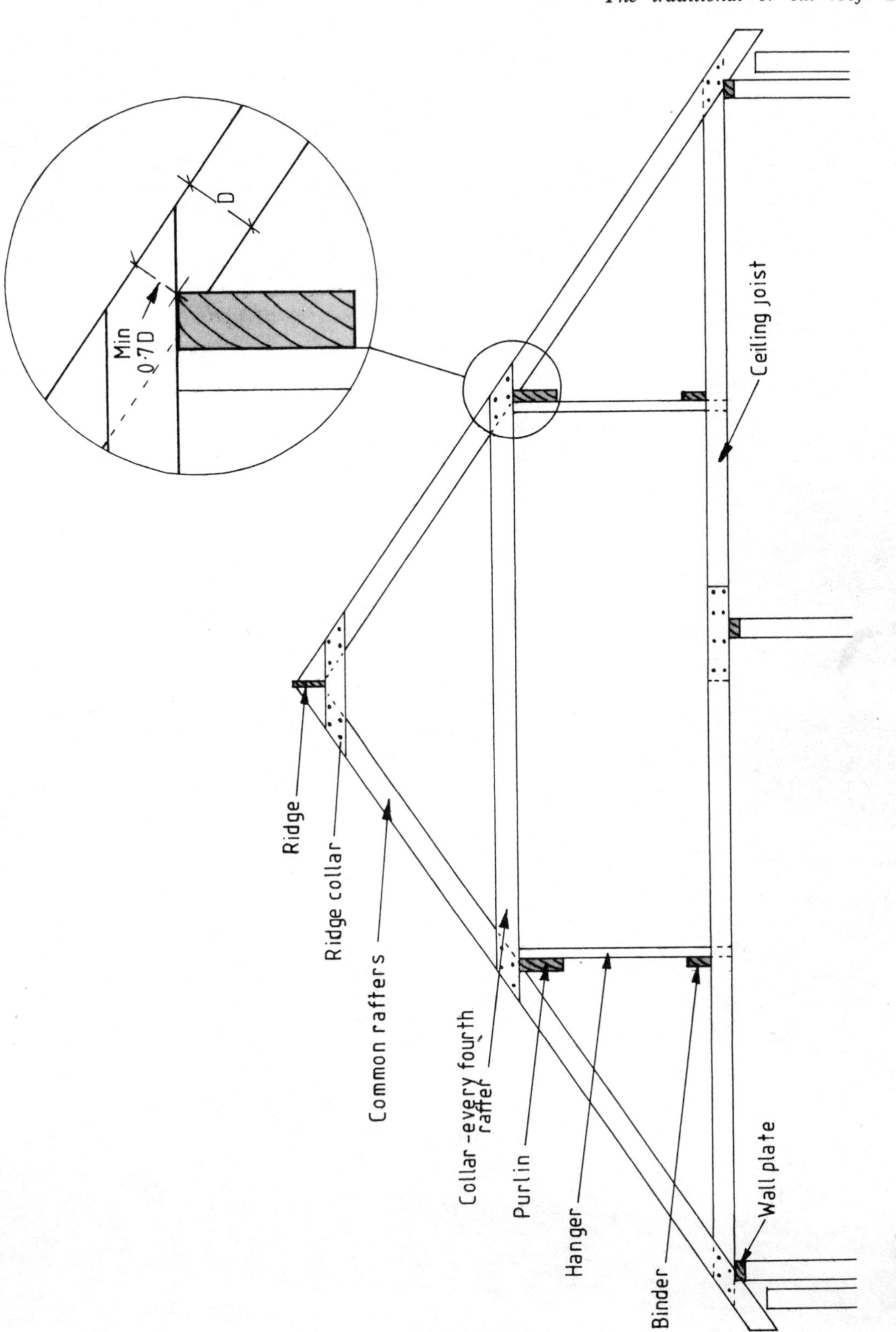

D
Min
0·7D
Ceiling joist
Ridge
Ridge collar
Common rafters
Collar - every fourth rafter
Purlin
Hanger
Binder
Wall plate

Fig. 3.2

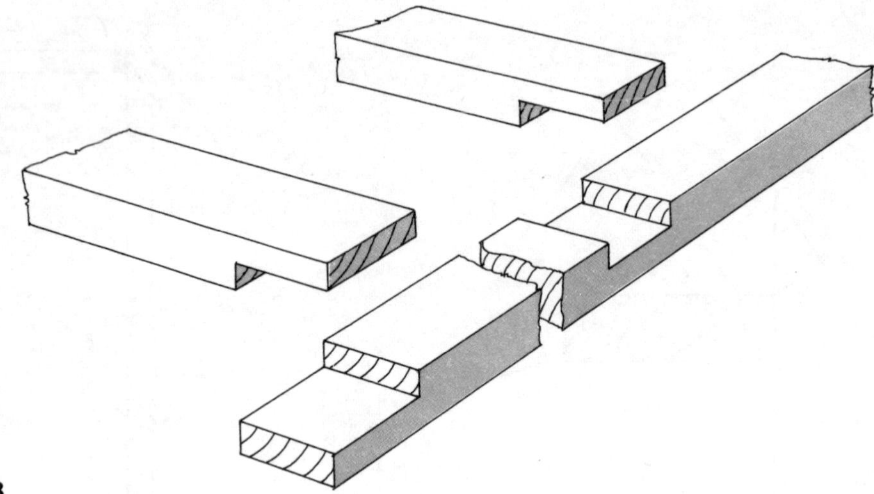

Fig. 3.3

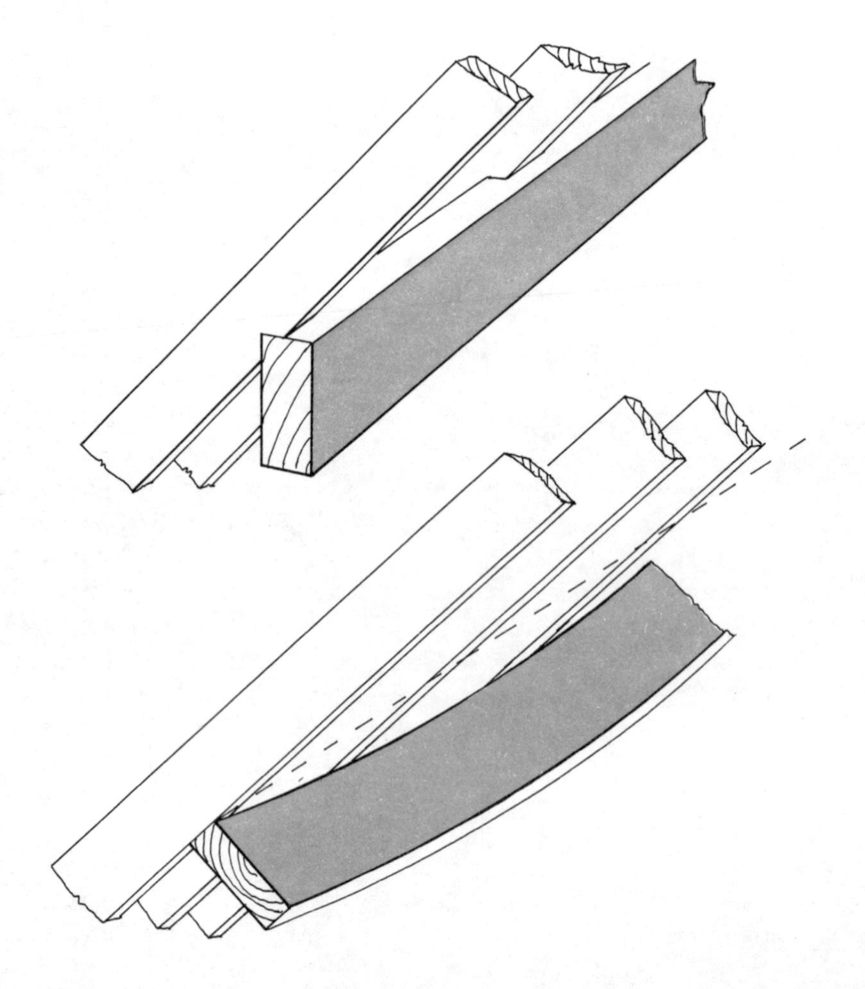

Fig. 3.4

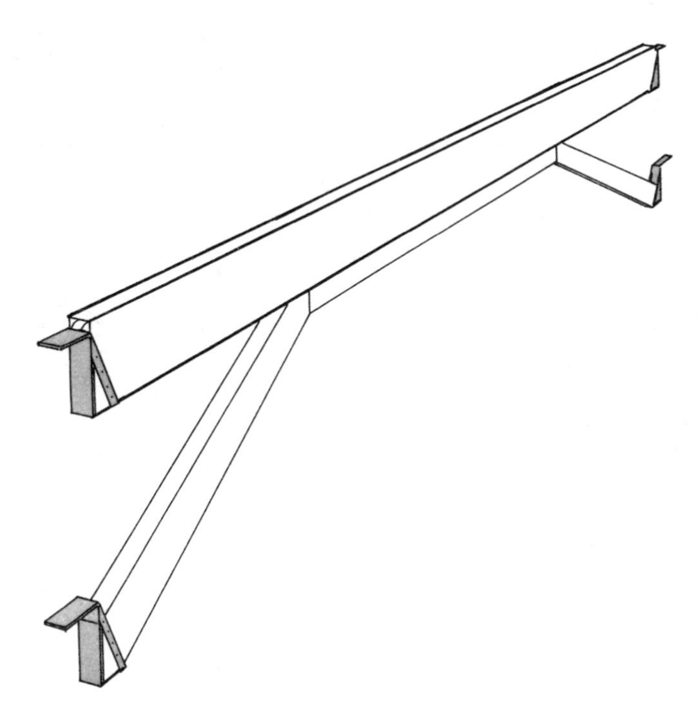

Fig. 3.5

purlin struts, as illustrated in fig. 3.5. The latter is to be preferred for, although slightly more labour intensive, it does allow ultimate economy in timber section and the struts give a stabilising effect to the walls supporting the purlins.

One final point on purlins, care should be taken with regard to the Fire Regulations when building purlins in to dividing or party walls between terraced dwellings. Unfortunately the approving authorities vary somewhat from area to area in their approach to timber built in to what are essentially fire walls between the dwellings, some allowing timber to be built in provided there is a positive fire break between the ends of the purlins, whilst others simply do not allow timber to be built in at all. In such cases built-in steel shoes will be necessary as indicated in fig. 3.5.

On longer spans of purlins it may be necessary to use prefabricated beams, these being dealt with in more detail under the attic roof solution later in this chapter, the beams themselves being illustrated in fig. 3.13. An alternative solution on long spans is of course to provide an intermediate support for the purlin by means of a post which, in turn, is directly supported from a structural wall below.

Rafters

Little needs to be said about the common rafters as these can be simply designed from the span tables. However on a very long roof slope it may not be practical to obtain timbers in one continuous length. On roof slopes in excess of 4.8 m a second purlin should be considered as illustrated in fig. 3.6. A collar should be fitted on every pair of

Fig. 3.6

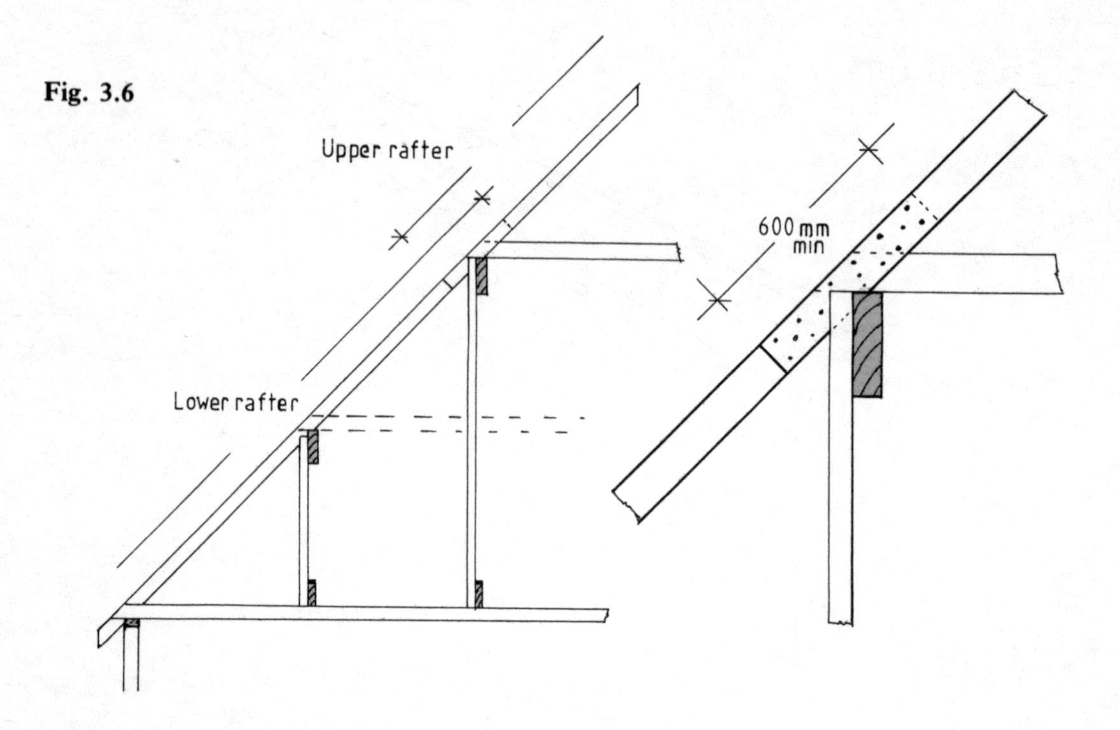

rafters immediately beneath the ridge, and a further collar should be fitted to every other pair of rafters immediately above the purlin position.

Ceiling joist

Even on most domestic roof spans it will be impractical to obtain a ceiling joist member in one length. It will also be necessary, unless very large ceilings joists are used, to support the ceiling joist at some point along its length, by suspending it from the structure above and/or on a structural wall below. Fig. 3.2 shows the typical solution using hangers and binders. More information on a typical hanger and binder combination for supporting the ceiling joists can be found in chapter four.

Connections

All timbers on this roof construction would normally be simply nailed together using 75 mm and 100 mm long galvanised wire nails. Whilst a simple 'tosh' or 'skew' nail (i.e. a nail driven at an angle through one piece of timber into its supporting timber) will be adequate on the rafter to plate purlin and ridge connection, the collar to rafter and ceiling tie to rafter connection should be made with three or five nails depending on the size of the individual members and taking care not to nail too near the ends of the timbers, thus avoiding splitting.

THE HIP ROOF

A simple hip roof is illustrated in fig. 3.7. Whilst the wall plate is the main support for such a roof, the main problem of support arises from the lack of a gable end from which to support the purlin in the hip area. The mechanics of load distribution within the hip area seems to be open to debate. It is quite clear however, that the majority of the load is transmitted directly to the wall plate with the symmetry of the jack rafters leaning against each other either side of the hip rafter, tending to provide a self supporting structure. Certainly on small span roofs where no purlin is required this would be the case. On larger spans however, where a purlin is required on the longer jack rafters, a more sophisticated solution must be found.

The plate

The wall plate need only be a perfectly standard timber section, but with thrust from the hip rafter being resolved at the external angle of the wall plate, it is common to fit a tie across the corner. A more sophisticated corner joint used on some older buildings is illustrated in fig. 3.8, but it would not generally be necessary for the size of the structure normally encountered on dwellings. It does however illustrate what was found necessary to contain the thrust from larger hip rafters.

The purlin

The need to support the purlin in the hip area has been mentioned above. One solution is to identify a suitable wall immediately beneath the hip area and use this for a support for the purlin. It is however more likely that the wall will be slightly outside the hip area as illustrated in fig. 3.7, thus necessitating a degree of cantilevering of the purlin itself. The post should be at least twice the thickness of the purlin it supports enabling a halved joint to be cut at the top thus allowing the purlin to fit squarely on a timber joint and not relying simply on nails. The supported purlin running across the hip end between the ends of the main purlins should again be halved onto the ends of the main purlin, thus providing a positive support.

Rafters

Rafters can be designed as for the more simple roof described earlier – the jack rafters will maintain the same cross-section, be birdsmouthed over the plate and nailed either side of the hip rafter. The angle of cut on the rafter abutting the hip rafter is what is known as a compound angle and this, like many other of the angles necessary on timbers in roof structures, can be calculated from the 'carpenter's square' or by reference to such specialist sets of tables as can be found in the *Roofing Ready Reckoner* (see the bibliography).

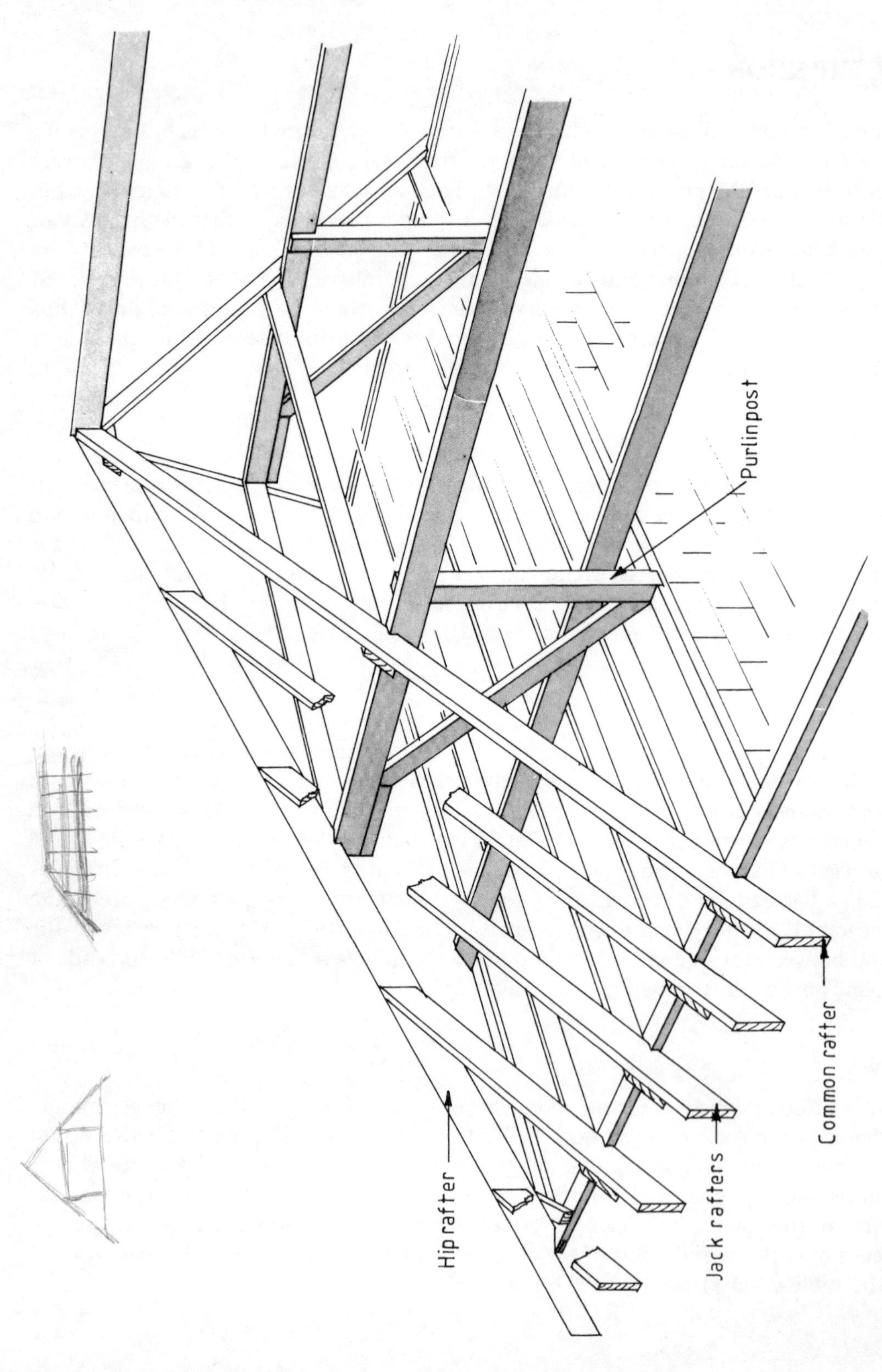

Fig. 3.7

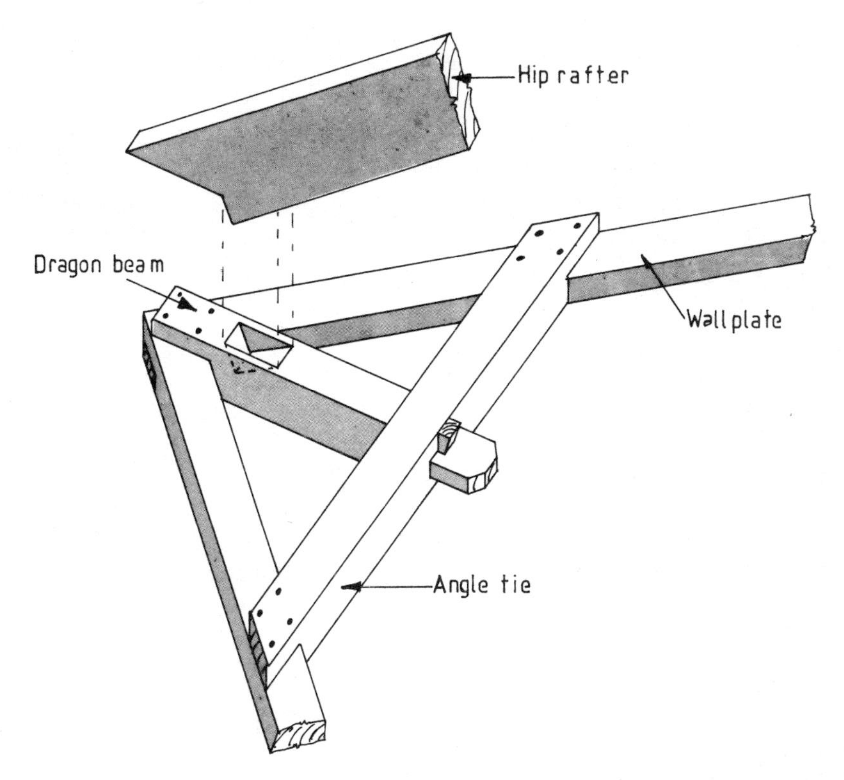

Fig. 3.8

Ceiling joist

In the writer's opinion it is important to maintain the tying effect of the ceiling joist within the hip area and for this reason the ceiling joists should span in the direction indicated in fig. 3.7. To do this it will be necessary to maintain the support by the binder within the hip area, and whilst the binder can be supported from the cantilevered purlin, it may be more prudent to have a separate binder beam supported at its ends, on the extreme end of the roof at the wall plate, and on an internal supporting partition wall.

THE VALLEY STRUCTURE

The valley is a very common feature of domestic roof structures. Some of the common roof shapes are illustrated in fig. 3.9 with full valleys, i.e. a valley running from eaves to ridge in parts A and B, with shorter valleys in parts C and D.

If one considers the hip to be an external mitre of the roof, then the valley is an internal mitre. The easiest way to consider a structural solution is to imagine one roof being wholly or partially imposed upon the other, and this is most easily illustrated in fig. 3.9B where the top part of the roof can be imagined to run through undisturbed as

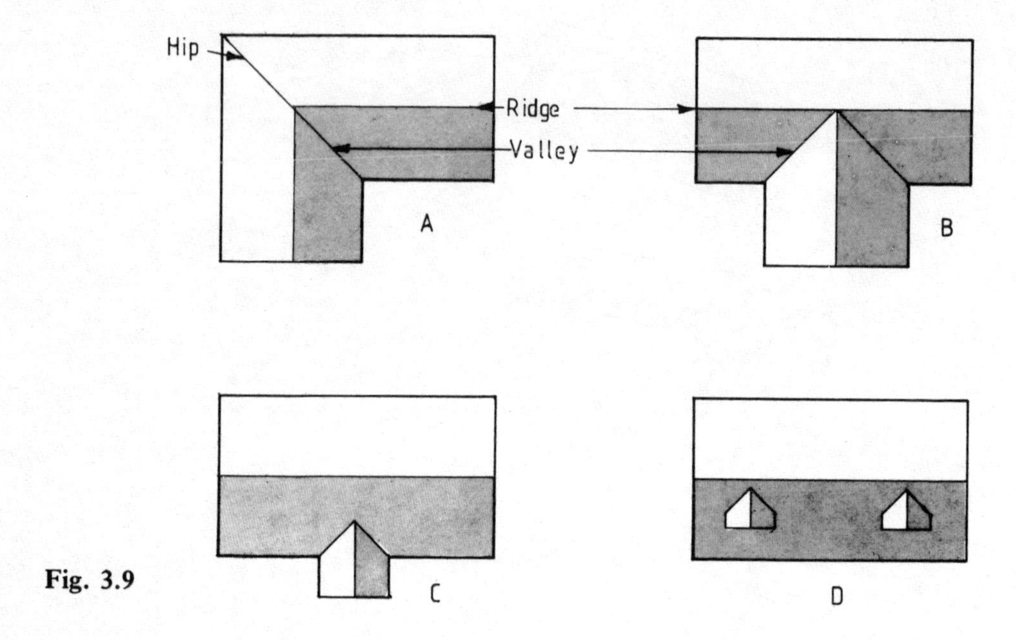

Fig. 3.9

a normal gable to gable roof, with the leg of the T imposed upon it. Figs 3.9A-D show other valley situations.

Fig. 3.10 shows a typical solution with valley jack rafters imposed upon a valley board which is itself supported by the main roof common rafters. The solution does of course assume that the common rafters will themselves be supported by a wall plate or beam immediately beneath the valley area. The alternative solution where this support is not provided is shown later in the solution for valleys on attic roof structures.

On valleys where the rafter length itself needs purlin support, the purlin in the valley area should be arranged at the same height as the purlin in the main roof. Furthermore it should be supported where it passes over the wall plate line of the main roof by a post, and by a steel hanger or shoe where it adjoins the main roof purlin. Fig. 3.11 illustrates the purlin connections.

ATTIC ROOFS

The attic or room-in-the-roof construction has become increasingly popular in recent years with the tendency for planners to seek steeper pitched roofs, particularly in rural areas. The house style created is often referred to as a 'chalet'. On a normal two-storey house with a roof pitch of about 45°, the volume enclosed by the roof is approximately 40% of the volume of the two storeys below, and on a single-storey building this proportion goes up to almost 80%. It therefore makes sense to attempt to use the extra enclosed space provided by the roof structure.

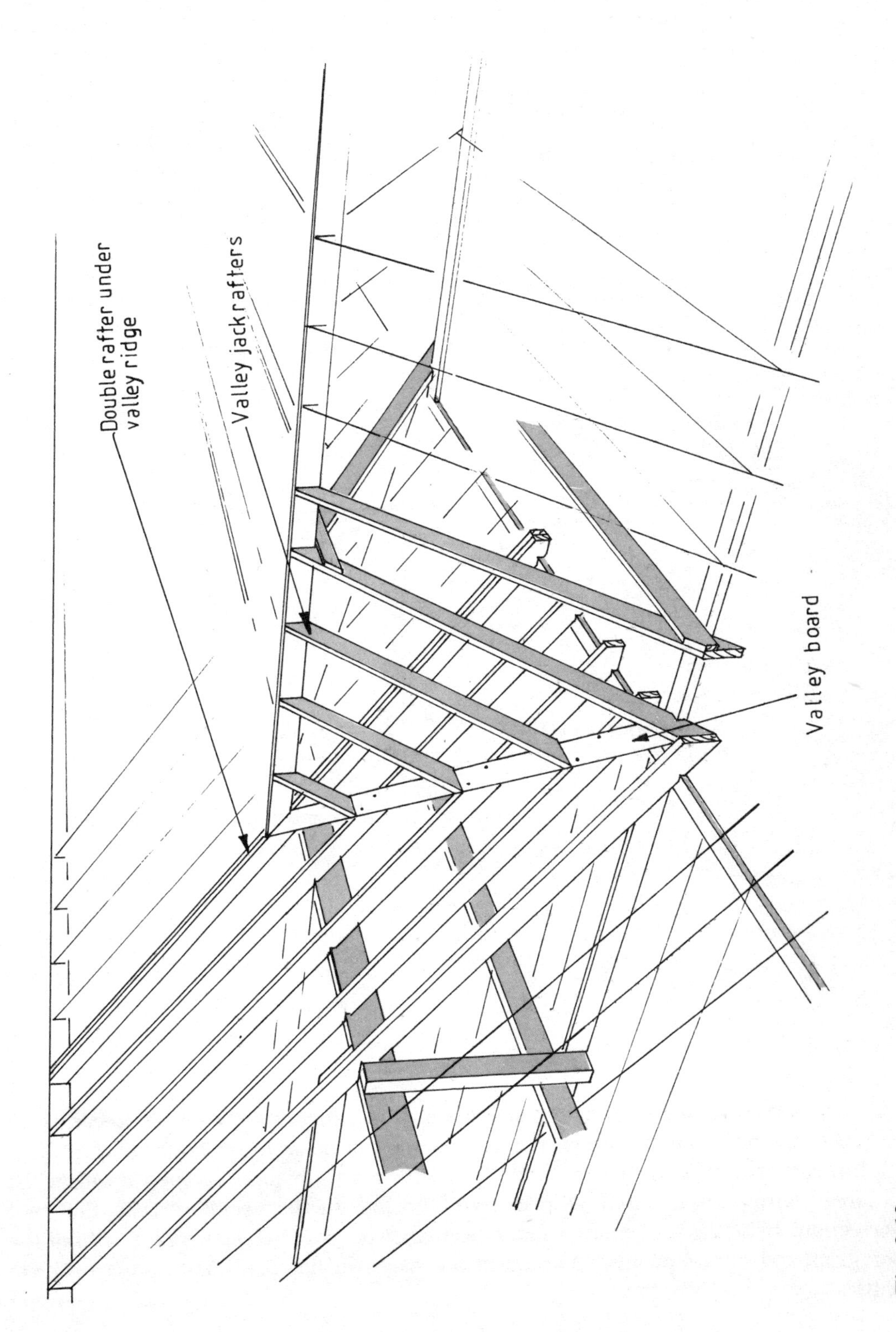
Double rafter under valley ridge
Valley jack rafters
Valley board

Fig. 3.10

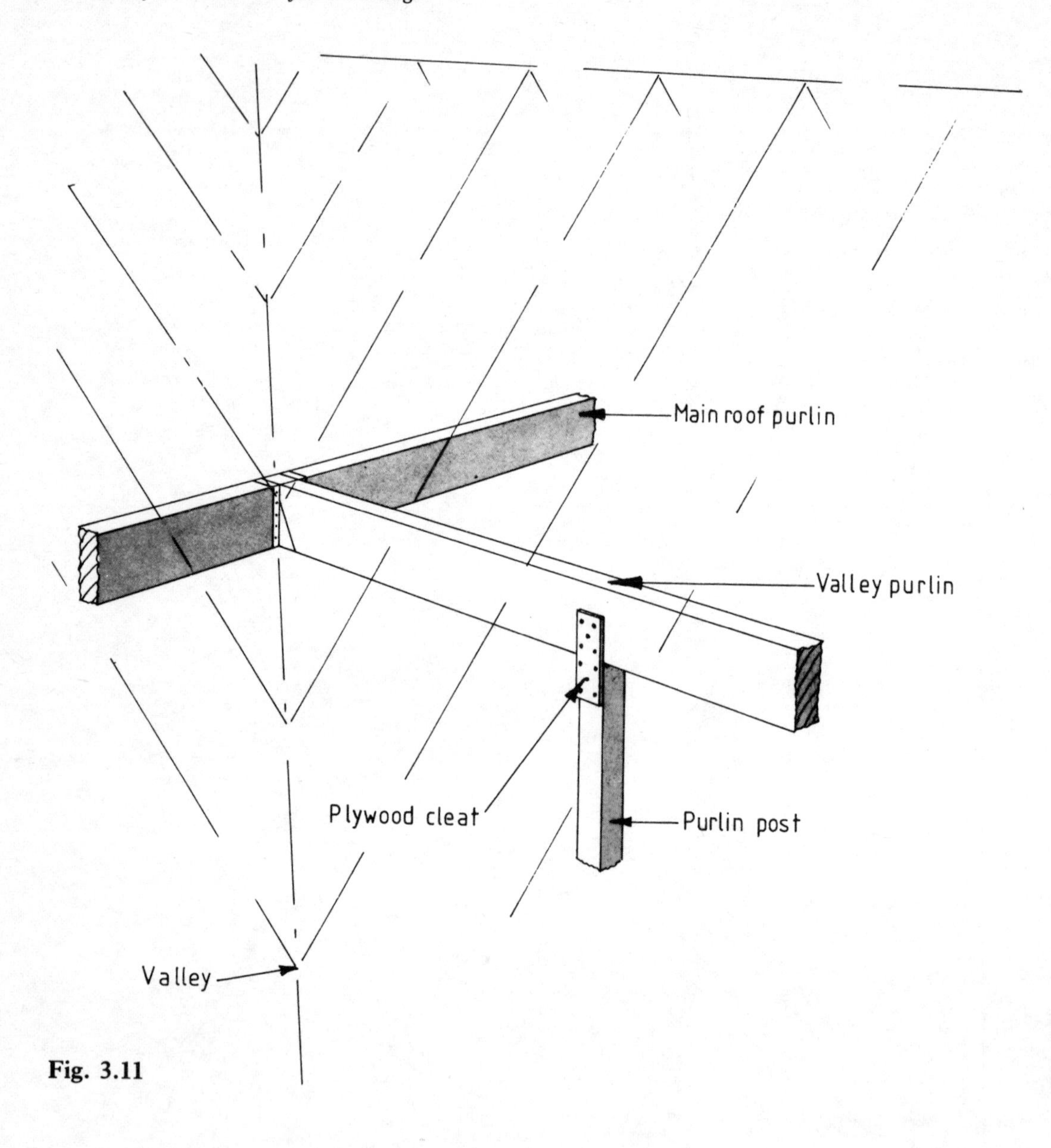

Fig. 3.11

An attic roof structure with gables at both ends is comparatively simple to construct using purlins and common rafters, this roof shape can be seen in Chapter 7, fig. 7.22. The hip end attic however poses some more difficult problems of support within the hip area, with the L or T-shaped roofs posing further problems at the roof intersection. In all cases careful consideration must be given to the support both of the floor joists and of the purlins. The question marks in fig. 3.12 show these problem points.

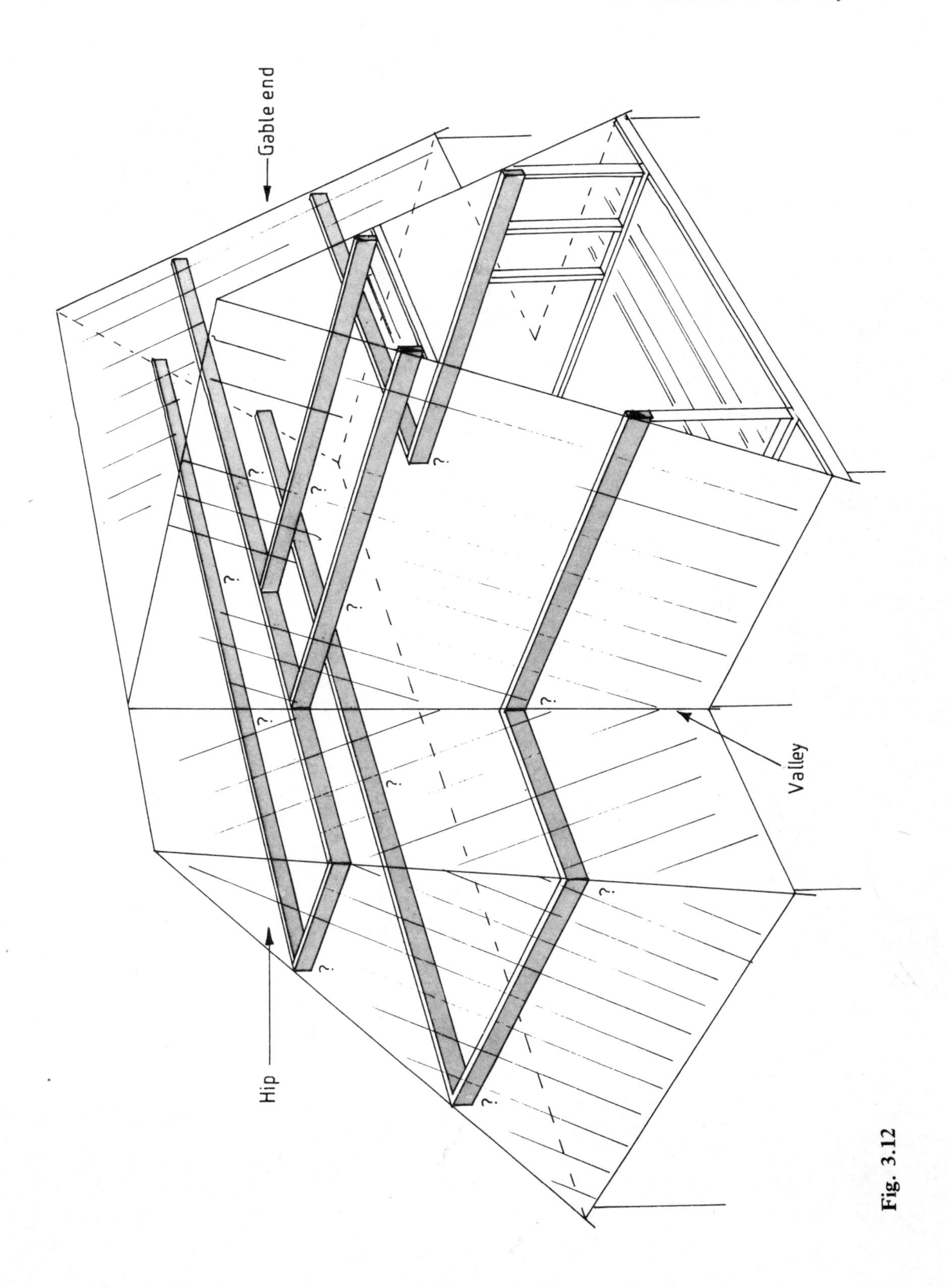

Fig. 3.12

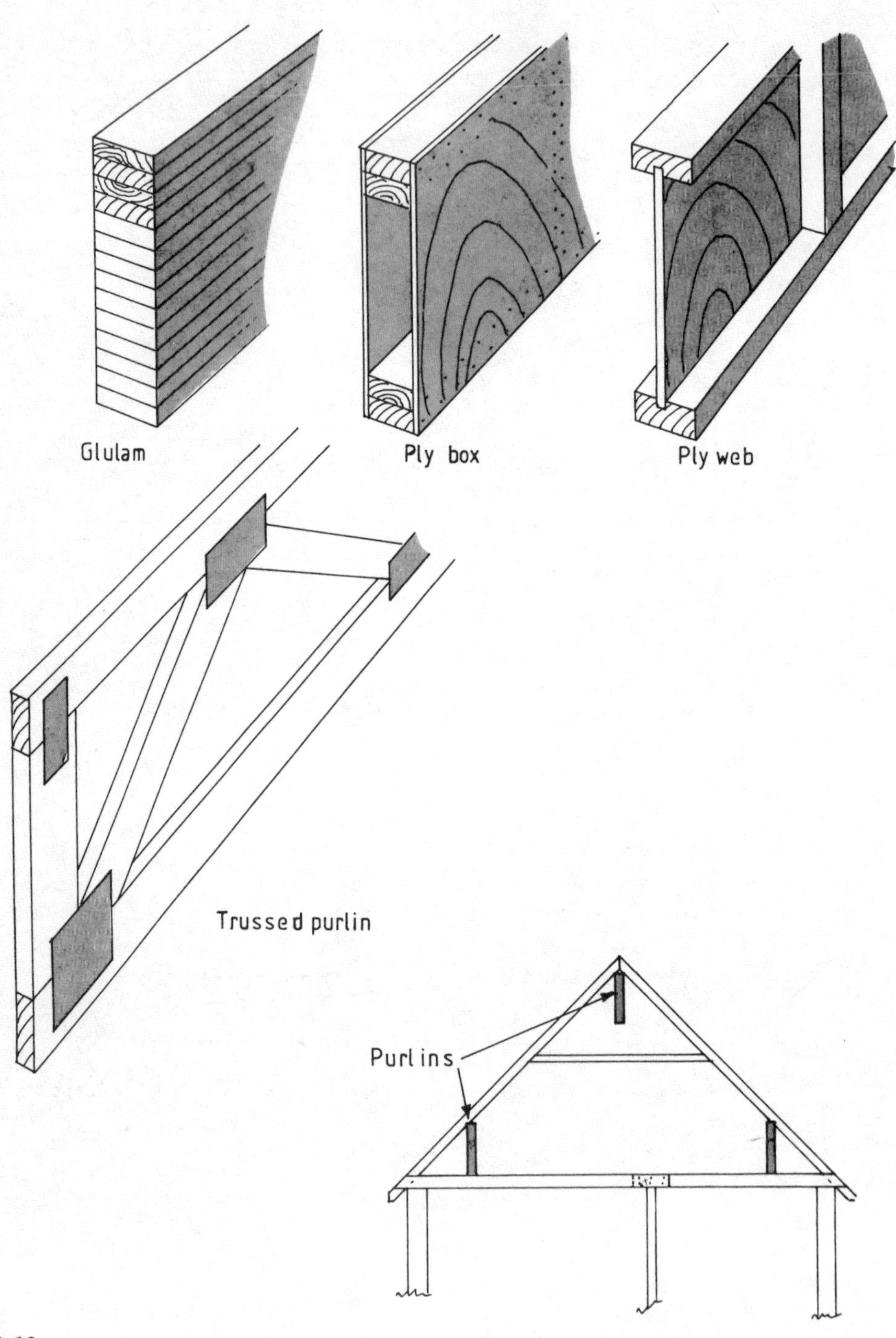
Glulam
Ply box
Ply web
Trussed purlin
Purlins
Typical large purlin roof

Fig. 3.13

The simple attic

The floor joists will seldom be able to span from external wall to external wall and will therefore need some internal support either in the form of an internal wall, or a beam. Similarly the purlin is unlikely to be able to span from one gable to the other without internal support. Whilst the purlin problem can be eased by using beams capable of larger spans it is likely that some internal support will still be required. Examples of larger span beams are indicated in fig. 3.13, giving spans of up to about 9 m without internal support.

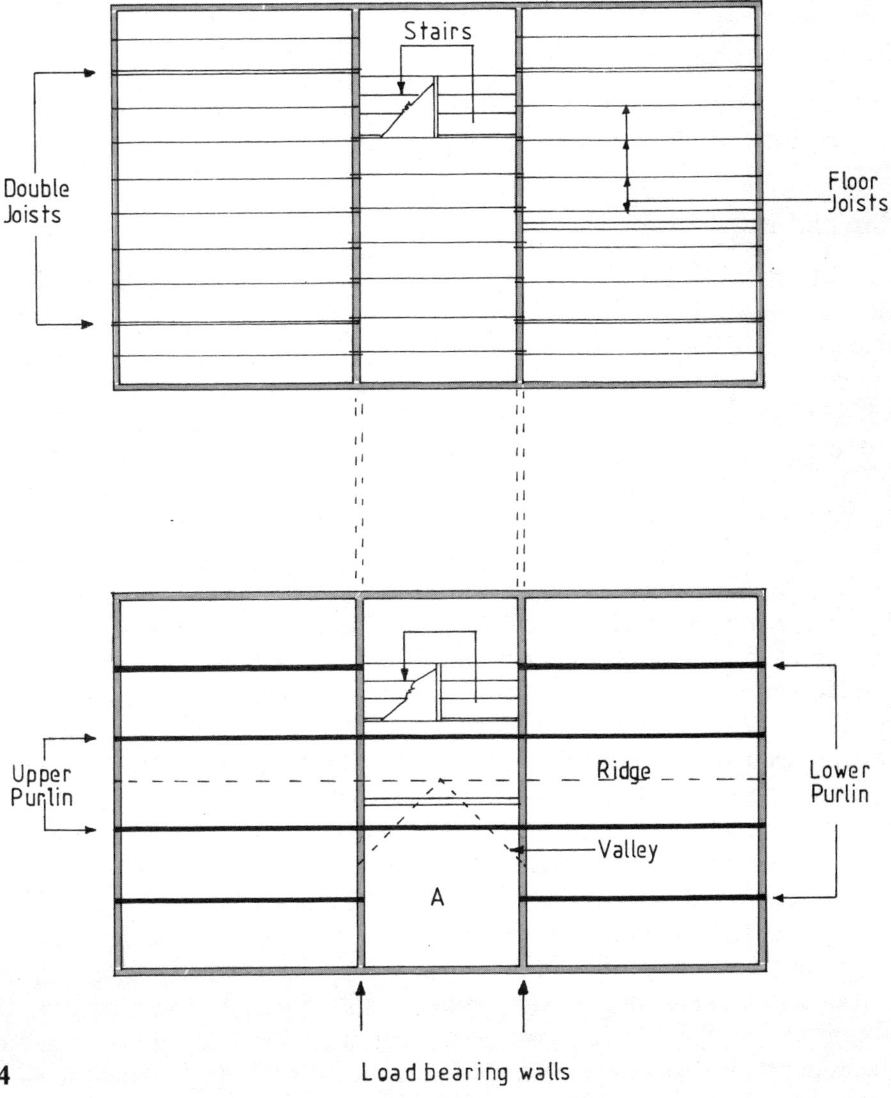

Fig. 3.14

Clearly it can be seen that in considering a common and purlin attic roof, the design must extend all the way down to the foundations below the load bearing internal walls. Support for the purlins by internal walls within the attic area will mean support in the form of beams or walls on the lower floor, thus to a certain extent controlling the room layout. A simple example of this is indicated in fig. 3.14, based on the layout of fig. 3.16.

It is therefore almost impossible to provide a solution to an attic construction without knowing precisely the room layout, and which of those room walls is capable of carrying load. To aid construction on site, the design should have:

(1) Common floor joist depth
(2) Common rafter depth
(3) Common purlin depth
(4) Common purlin lines

Structural economy can be achieved by varying the thickness of individual members and/or their grade stress, and the spacing of structural members.

The hip and valley attic

Having highlighted the need for both floor joist and purlin support, on a simple attic, consideration must now be given to the solution of the more complicated attic roof as illustrated in fig. 3.12.

It can be assumed that the floor joists will generally be supported either by beams or by load bearing walls below. In some instances these joists may be doubled to form a beam, or a separate beam is inserted within the floor which will itself be strong enough not only to carry the floor but also to support the purlin above. This however generally applies only to the lower purlin, i.e. the one at the attic wall to the sloping ceiling junction, and not to the upper purlin. The latter is more likely to be supported by internal load bearing walls.

Reference should be made to fig. 3.15. At the gable end both sets of purlins can simply be built-in in the usual manner, with the floor joists supported by being built-in at the gable and either on a load bearing room dividing wall internally or a beam as illustrated.

At the hip end the construction is more complicated, but again floor joists can be supported on the wall plate of the external wall and on either a beam or internal load bearing partition. The lower purlin at the end of the hip can be either supported at its ends, strutted up from a specifically designed double joist or beam below, or it can become little more than a wall plate supported by the end wall studs from floor joists below. In the latter situation of course the floor joists must be designed to carry this additional load. The side purlins in the hip area are most likely to become in effect wall plates, themselves being supported by studs at 400 mm or 600 mm centres, the studs being supported by the double joist or beam below. If the side wall is to be omitted to give maximum floor space then the purlin must be capable of spanning the room width and will need to be supported by room dividing walls.

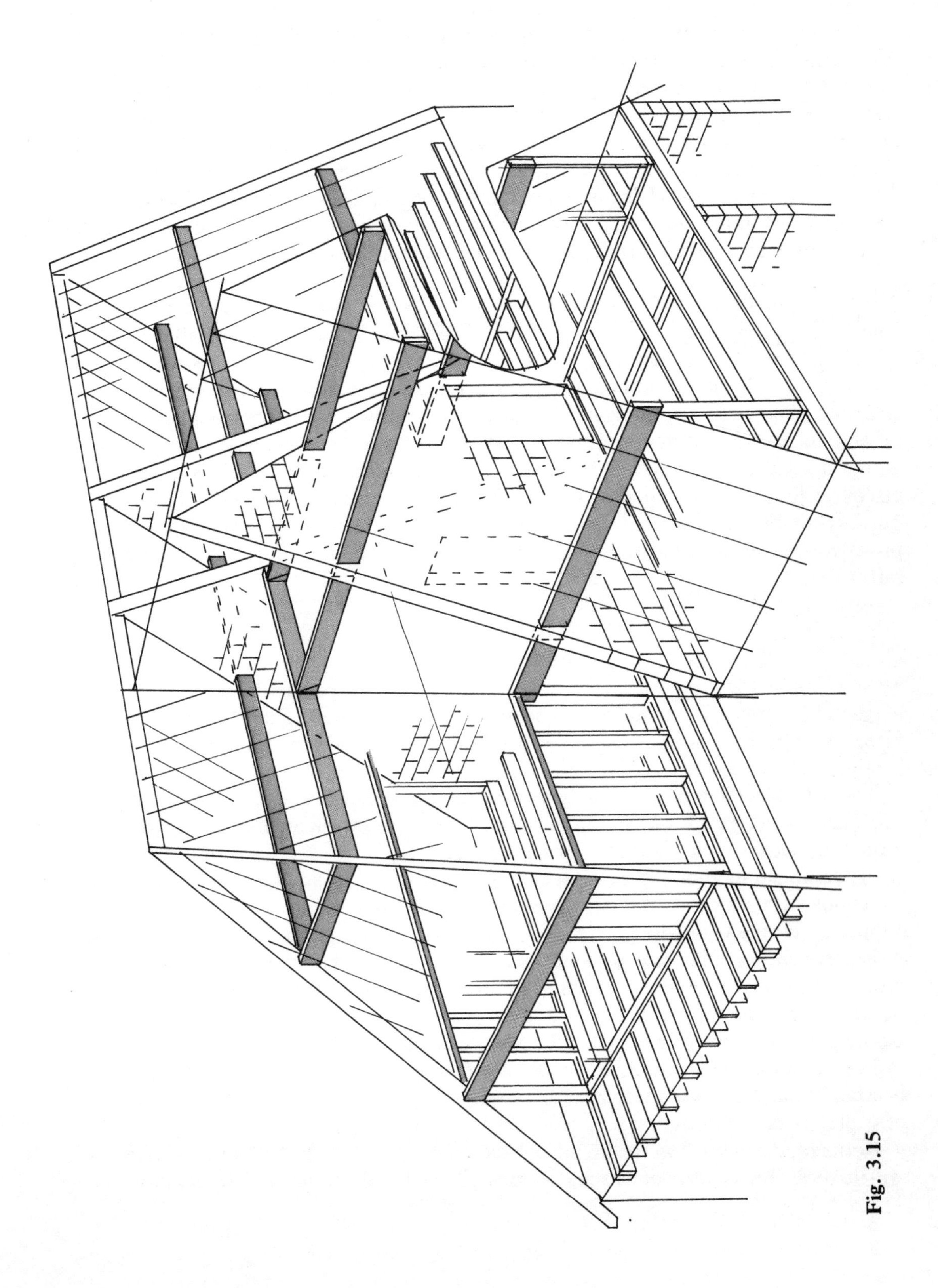

Fig. 3.15

The upper purlin will generally span from gable to internal load bearing walls and then from internal load bearing walls to further load bearing internal walls as illustrated. However in the hip area it is not possible to provide an adequate cantilever length, bearing in mind minimum room size requirements, and it is certainly not possible to provide a strut as illustrated in fig. 3.7. The purlin at the hip rafter junction therefore may well have to be supported by the hip rafter itself, thus making the hip rafter a major structural element. Where this is the case it will not be possible to design the roof structurally from fig. 3.1 and design advice must be sought. For the purposes of the illustration such a structural rafter has been assumed.

The valley intersection
The lower purlin in fig. 3.15 is supported by the internal load bearing wall and will to a certain extent cantilever into the valley area. However additional support may be provided by a post down onto the beam provided in the floor of the main roof structure. The upper purlin again supported by the internal load bearing wall will not be able to provide a full cantilever and must therefore be connected into the purlin within the main roof structure. The connection will normally be by a steel shoe and is of course following a similar structural layout to that indicated in fig. 3.11, with the purlin post being replaced by the internal wall. Rafters, collars and ridge may be provided bearing in mind the considerations described earlier, especially concerning rafter length.

Attic dormers

Most attic roofs will be fitted with at least one dormer to provide both increased full-height room area within the roof, and light and ventilation. Roof windows may be fitted to provide light and ventilation but they do not add significantly to the full-height room area.

The first dormer type construction to be considered is that illustrated in fig. 3.16, the main structure of which can be seen to be brickwork continued up from the structure below. Such a structure would provide the roof over the area marked A in the lower illustration on fig. 3.14. It can be seen that this follows a conventional valley situation, with valley jack rafters supported by a ridge onto a valley board in turn supported by the common rafters of the main roof. The common rafters in this case, rather than being supported by a wall plate, are supported by the upper purlin of the attic roof. The true dormer window, examples of which are illustrated in fig. 3.17, occur within the roof slope, i.e. not starting at the eaves and probably finishing well before the main roof ridge. Fig. 3.9D shows two such dormers.

The additional load imposed by the dormer on the main roof is relatively small, bearing in mind that the most significant weight on a roof is the tiles, and this tile area of course is not increased. The additional load must be allowed for.

Dormers, depending on the architecture of the area in which the house is to be built, will have a variety of roof shapes. Whilst fig. 3.17 illustrates a flat, mono-pitch or

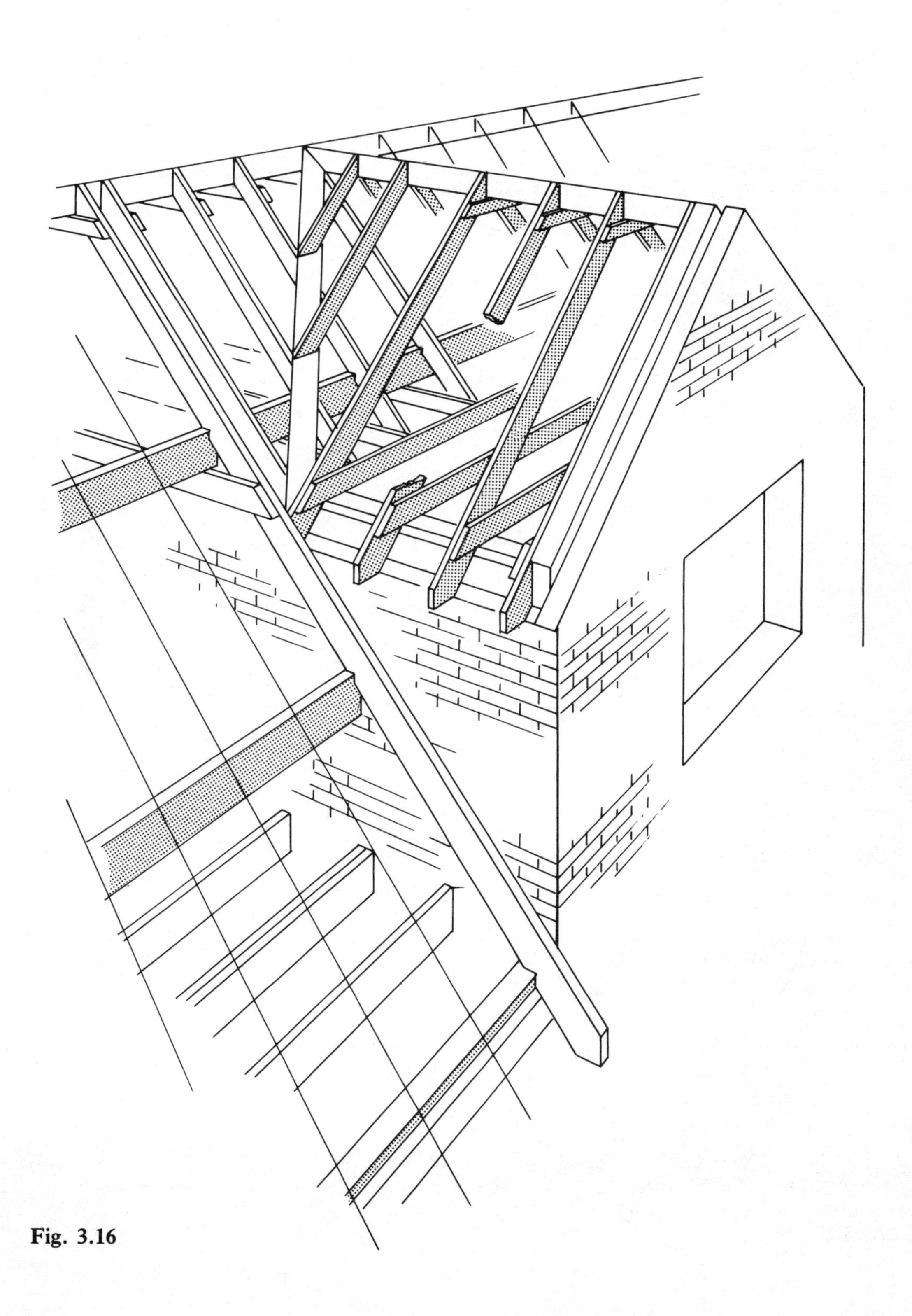

Fig. 3.16

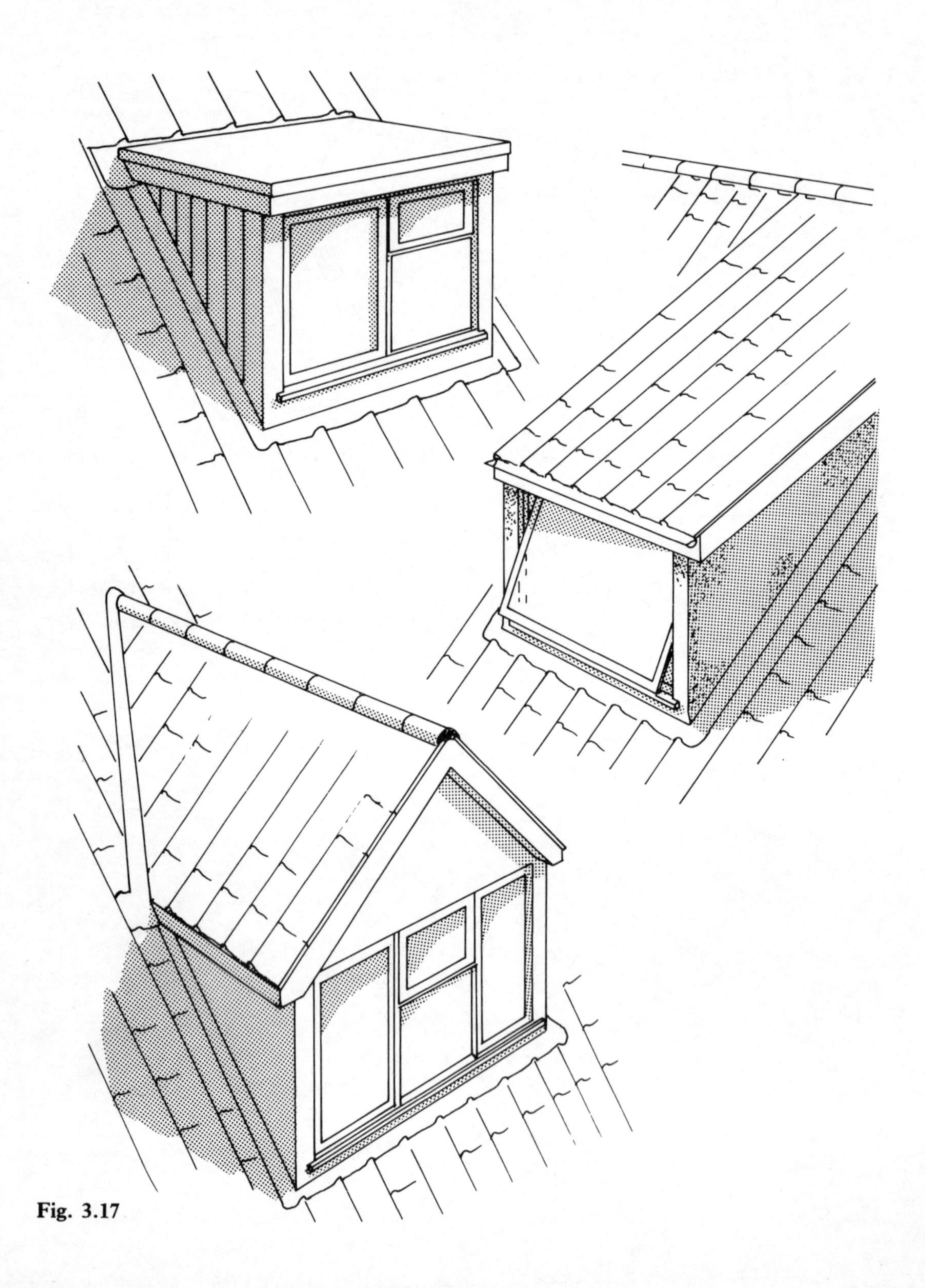

Fig. 3.17

'catslide' and a conventional symmetrically pitched roof with gable ends, hip end dormers are not uncommon.

Dormer framework
The construction of the dormer is relatively simple, the critical part of the design is the forming of an adequately strong framed opening within the main roof. In forming the hole for the dormer, the continuity of roof slope support is removed and provision must be made to carry the load both above and below the opening formed. If an upper and lower purlin are used in the attic structure, then these members may be used to support the rafters both above and below the opening. The perimeter of the hole formed in the main roof will provide the foundation for the dormer framework itself. Fig. 3.18 illustrates the method of imposing simple dormer framework onto a trimmed opening. Fig. 3.18a illustrates the rules for trimmer numbers.

ROOF LIGHTS AND ROOF WINDOWS

Roof lights generally will require much smaller openings within the main roof structure than a dormer described above, therefore a similar method to that illustrated in fig. 3.18 will be more than adequate. However, as the roof lights may not extend up the roof slope the full distance between two purlins, separate secondary purlins or trimmers may have to be introduced. If this is the case the rafters onto which the trimmers are fixed must be reinforced by attaching an additional rafter to each side of the opening, these additional rafters extending from the lower to the upper purlin (see fig. 7.26).

The term 'roof window' is a term recently introduced to the building industry to describe what is in effect an opening roof light. The roof light is normally fixed, of course, and provides only light to the building. Roof windows are most commonly of proprietary manufacture with the manufacturers providing detailed guidance on the method of fitting the roof window to both existing and new roof structures. Reference should be made to the manufacturer's instructions if such a roof window is to be fitted.

The advantage of the roof window over the dormer, if additional floor space is not the criterion, is that because the glazed area is angled directly at the sky, significantly more light is admitted to the room. The manufacturers of the proprietary roof windows also claim up to 70% saving in cost over a comparable dormer construction.

ADDITIONAL DESIGN CONSIDERATIONS

New houses constructed under the control of the *National House-building Council* must further conform to the requirements of the *Registered House-Builder's Handbook*, part II of which sets out the technical requirements for the design and construction of dwellings. The reader is directed particularly to the section on carpentry, section Ca,

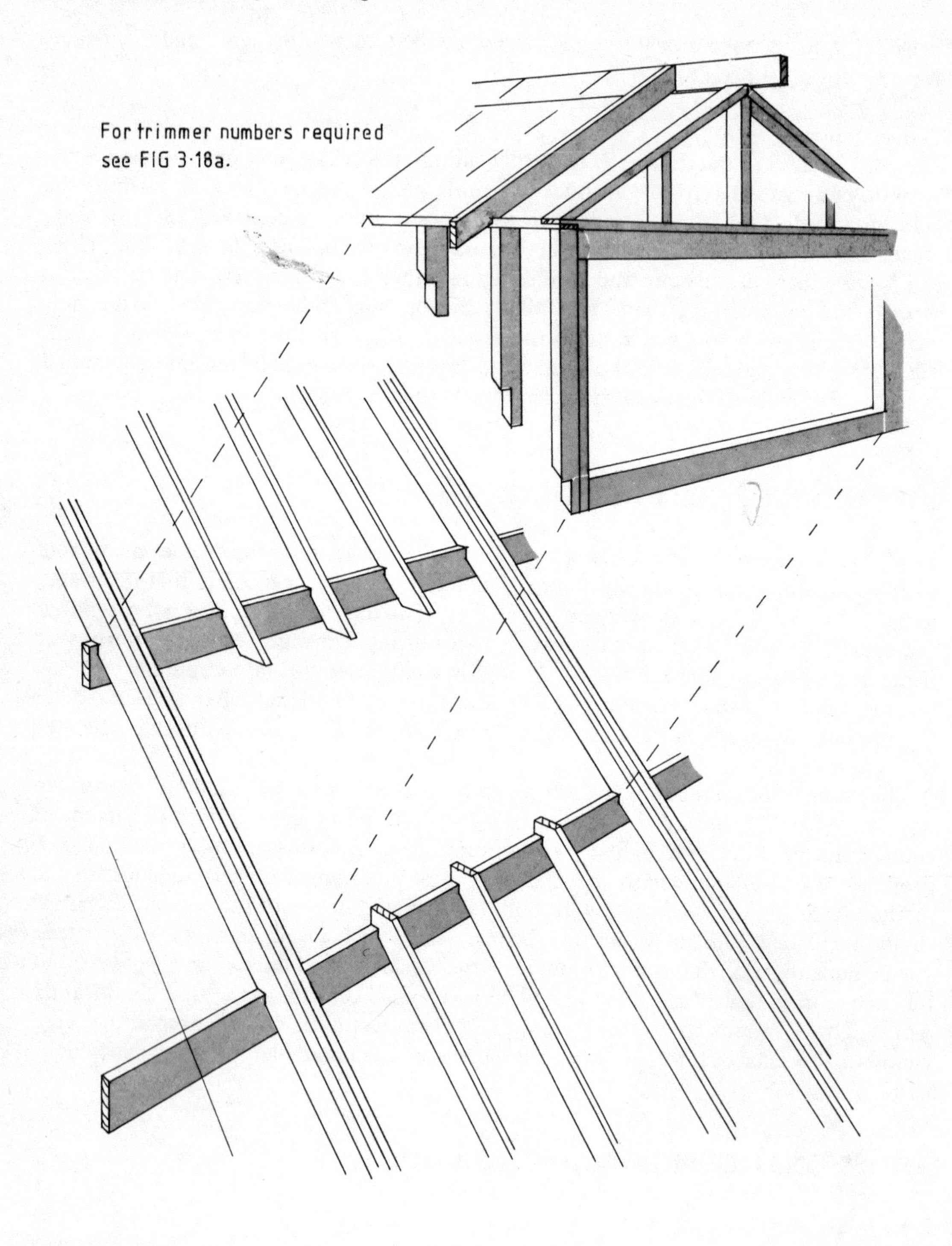

Fig. 3.18

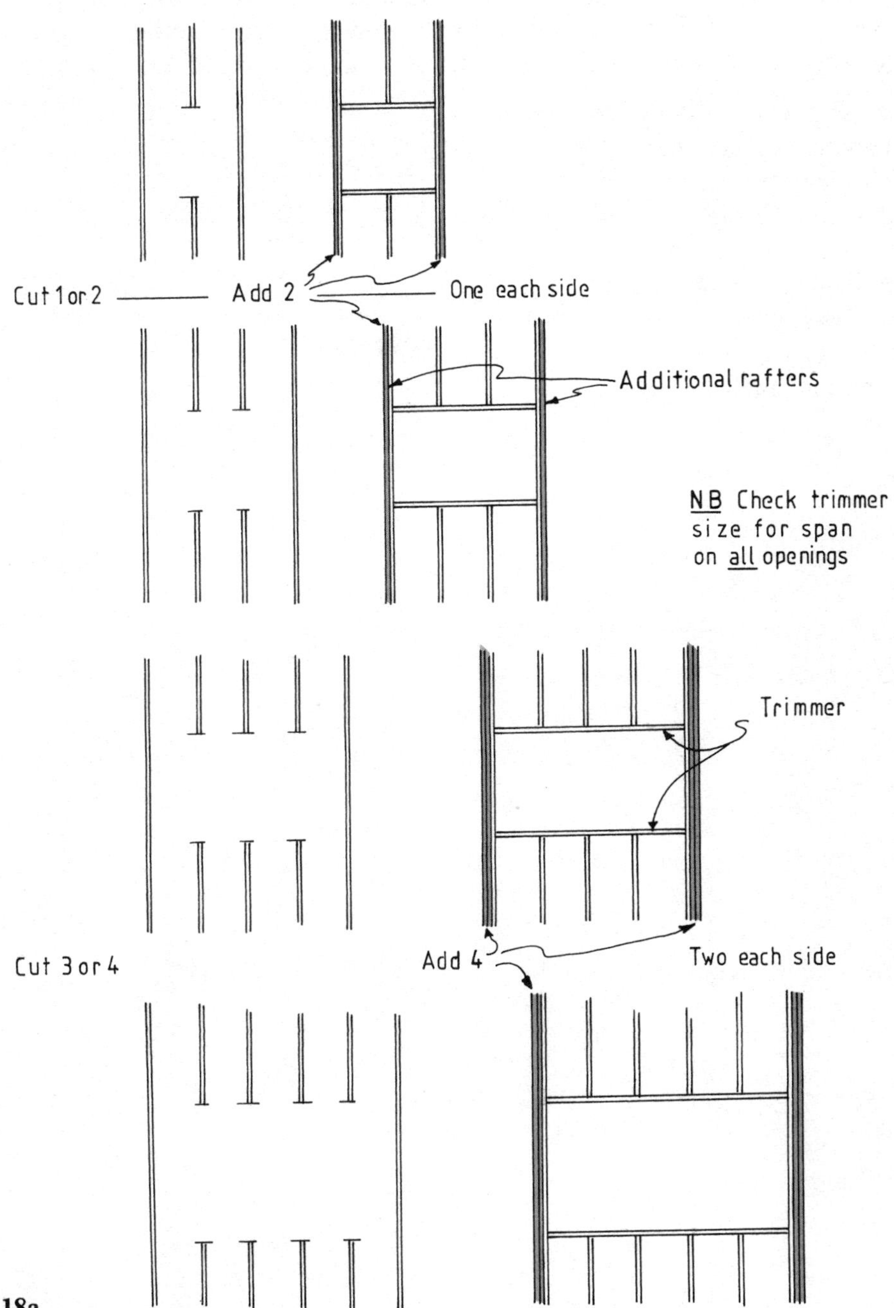
Cut 1 or 2
Add 2
One each side
Additional rafters
NB Check trimmer size for span on all openings
Trimmer
Cut 3 or 4
Add 4
Two each side

Fig. 3.18a

and also to the Schedule of Facilities and Services clauses S14 and S15 which concern thermal insulation and ventilation of roofs and access to lofts respectively.

Chapter 7 of this book concerns itself with the many other aspects of roof construction which apply to all roof forms. The items covered are as follows:

(1) Storage and handling of timber
(2) Preservative treatment
(3) Wall plates and straps
(4) Gable ends, straps and gable ladders
(5) Water tank platforms
(6) Ventilation of roof voids
(7) Roof bracing
(8) Eaves details
(9) Trimming small openings.

CHAPTER 4
Bolted truss roof construction

Roofs of all shapes and sizes can be structurally designed using bolt and connector jointed truss forms. Individually engineered designs for bolted trusses are used for public buildings where they are often left exposed as a feature of the design of the building. The bolted truss designs most frequently found in domestic dwellings are those prepared by TRADA, and it is these designs which will be dealt with in this chapter.

THE JOINTS

The bolt and connector roof structure is comprised of principal trusses spaced at centres dictated by the structural design. The principals in turn support purlins and common rafters, with binders supported by the principals, carrying the ceiling joists. The strength of the principal is quite naturally in its joints between the timber members, and therefore careful assembly of the principal truss is essential if the roof is to perform satisfactorily. Many such roof forms have been constructed on dwellings since the advent of the TRADA designs, and it is the sturdiness of the design, rather than high standards of workmanship, which has led to the method's success.

A simple bolt and connector joint is to be seen in fig. 1.11. This illustrates a double-sided toothed plate connector which is held in position (having first been embedded in the meeting timber surfaces) by a bolt with large load spreading washers. The alternative to the double-sided toothed plate, is the 'split-ring' connector and this will be dealt with later. It is essential that the teeth of the timber connector be fully bedded into the timber surfaces if full design strength is to be developed. It is the connector and not the bolt which is principal strength of the joint, the bolt mainly being there to hold pieces of timber together and ensure embedment of the connector is maintained. It is therefore essential that the nuts on the bolts are tight and some retightening may be necessary once the truss form has been installed as timber shrinkage may have occurred. It is quite common to have two or more connectors at a joint and the connectors themselves may vary in size and shape depending upon the design specification.

TRUSS ASSEMBLY

To understand the assembly sequence about to be described, reference should be made to fig. 4.4 which illustrates a typical connector jointed truss. For the purposes of the assembly sequence it will be assumed that all joints are made with double-sided toothed plate connectors.

Timber for one truss should be selected, cut to size, laid out on a flat surface into the precise truss shape required and all joints clamped together. Dimensions and pitch angle must now be checked and, if all are found correct, one can proceed to mark out the bolt centres and drill with a bit diameter not greater than 1.5 mm larger than the bolt diameter specified in the design. Care must be taken to maintain the drill square in all directions with the timber surface. If more than one truss assembly of the shape laid out is to be constructed, then this first truss must now be unclamped and the components used as 'masters' for the members of the remaining trusses to be constructed – this will not only save labour but will ensure that all of the trusses are identical. At this stage, then, the masters should be used to cut and drill all of the members required for the remaining trusses.

To assemble the first truss, lay out all the members and place a bolt through all of the joints except the first joint to be fitted with a connector. At this first joint (probably the rafter to ceiling tie joint) fit the specified connector and, using a special high tensile steel stud available from the connector suppliers, pass this through the bolt hole in the joint and place a large 100 mm square × 6 mm thick steel spreader washer on both sides of the joint. By turning the nut, bed the connector fully home. Considerable pressure is needed to bed the double-sided tooth plate timber connector and the mild steel bolts used to maintain the joint in the final assembly are not adequate for three reasons: firstly, the pressure required may well strip the threads from the mild steel bolts; secondly, if several connectors occur on the same bolt line there is unlikely to be adequate length of thread on a standard bolt to pull the joint down embedding all of the connectors (fig. 4.1 illustrates this). Thirdly, the smaller 50 mm × 50 mm × 3 mm thick mild steel washer used with the standard bolt will be inadequate to protect the timber surface from crushing and may itself be severely distorted. Where two or more sets of connectors occur, such as the ceiling tie to ridge strut joint on fig. 4.4, the appropriate number of high tensile studs will be required to pull the timbers down progressively if severe distortion of the timbers is to be avoided, as illustrated in fig. 4.1.

When the connectors are fully embedded the high tensile stud can be withdrawn leaving the joint generally well held together with the connectors. The standard bolt may now be inserted in the joint with the 50 mm square washers under both head and nut. The only pressure now required is that to maintain the joint in its bedded position. At this stage the nut should not be overduly tightened.

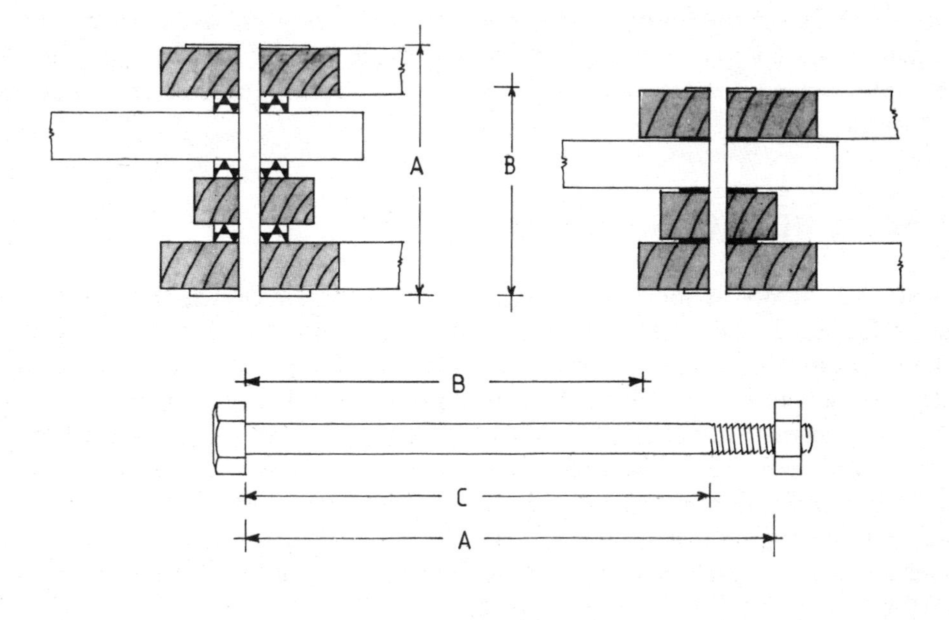

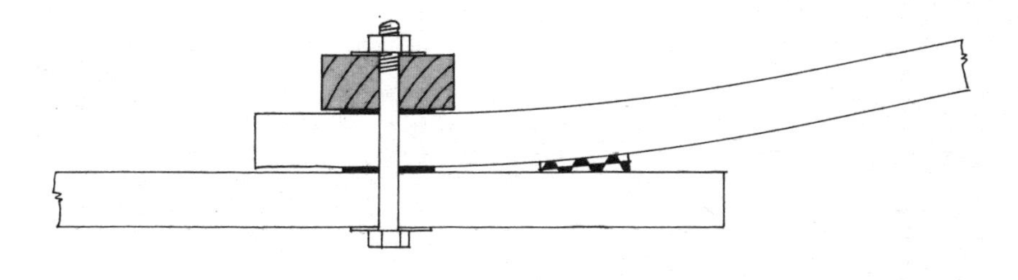

Fig. 4.1

The TRADA designs (referred to in chapter 1), show no collar under the ridge board, consequently for handling at the works (if the trusses are made off site), during transportation and also for on-site handling the truss is not jointed at the ridge. A temporary collar should therefore be fitted to avoid distortion of the truss and this collar should now be nailed into position on the rafters just below the ridge board line. On site, this will then aid the location of the ridge board during the roof construction. The space in the setting out of the truss to allow for the ridge board must not be varied because clearly it is vital to the overall geometry of the truss. A thinner ridge board will mean that the truss will sag and a thicker will result in the reverse.

The remainder of the joints in the truss can now be assembled as described above and when all are complete all the bolts should finally be tightened before the truss is moved. Subsequent trusses can now be assembled using the masters following the procedure described.

THE SPLIT-RING CONNECTOR

The split-ring connector joint is illustrated in fig. 4.2. The principle is similar to that in the double-sided tooth plate, in that the split-ring carries 75% of the load, the bolt 25% and maintains the timbers in position around the ring. The problem of embedment of the connector is solved by machining a circular groove equal to half of the width of the ring in each meeting timber surface. This groove is machined using a special 'dapping' tool available from the connector suppliers. Assembly of the split-ring connector jointed truss is similar to that for the design described above, but clearly all of the grooves for the rings must be machined before any final assembly takes place. The high tensile stud is clearly not required as the ring fits neatly into the groove allowing the final bolt to be placed through on the first assembly. The split-ring connector truss assembly is most likely to be carried out in a workshop where the dapping procedure can be carried out in fixed drilling jigs on benches. Dapping can be carried out on site but electric power to turn the dapper is almost essential, and a degree of skill is required in its accurate use. The bolt holes must be drilled first, the dapping tool is then located in the bolt hole to cut the groove for the connector.

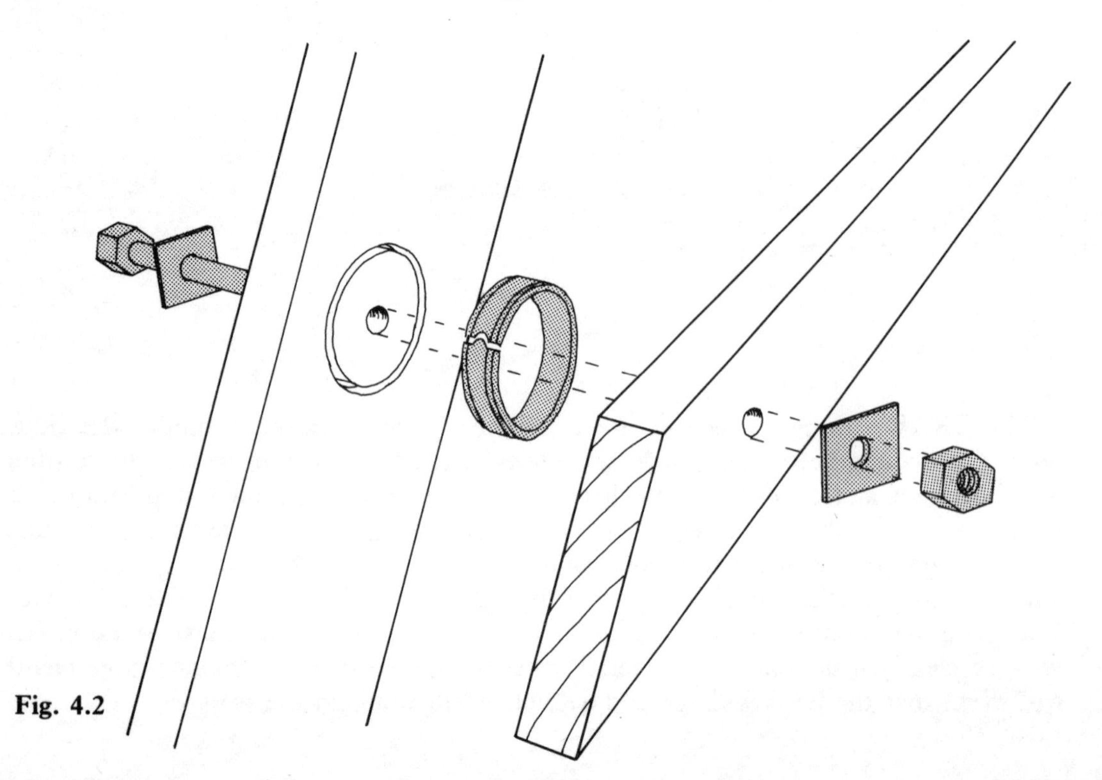

Fig. 4.2

Site assembly of bolt and connector jointed trusses using either connector form is of course a practical proposition, but factory assembly of the truss will generally result in a more accurate component simply because of the equipment, level floor area, dry conditions and specialist operatives used in the construction.

ACCURACY

Most bolt and connector trusses for domestic roof construction will be assembled using sawn timber. The tolerance on sawn timber in terms of over and under sizing, particularly in the larger dimension, is significant and for this reason the full size setting-out of the truss has been recommended in the assembly procedures described. The precise location for the bolt centres is essential to maintain the design edge and end distance for the timber conenctors themselves. Insufficient end distance may result in a timber connector shearing out a section of timber thus allowing the truss joint to fail. Great care must therefore be taken to follow the precise dimensions detailed on the engineer's drawing. Reference should be made in this respect to the dimensions indicated on the design illustrated in fig. 4.4.

STANDARD DESIGNS

TRADA have produced a range of roof design sheets, similar to that illustrated in fig. 4.4, for bolt and connector roof trusses for domestic and industrial use. Fig. 4.3 sets out the loading, pitch and span limitations for the designs available. All designs conform to BS 5268 part 2 1984, and are readily available from either the TRADA head office or one of their regional offices. Details of these can be found in the bibliography.

Although the designs cover a very wide range of span, tile weight and roof pitch, extension of the range may be acheived by reference to TRADA. The TRADA standard design sheet number 507 shows typical joint arrangements for the designs contained in sheets 500-511 inclusive.

THE ROOF CONSTRUCTION

The bolt and connector truss is generally used as a principal truss with common rafters and ceiling ties supported from it and on the wall plate. The designs referred to above are spaced at 1.8 m centres with the design of the purlin, binders, plate and ridge being given on each individual sheet. A study should be made of the design sheet illustrated in fig. 4.4. Further design and construction advice is given on the reverse of TRADA roof design sheets.

The construction of the roof itself, once the principal trusses have been produced, is quite straightforward, with many of the joints being nailed as with the 'traditional' roof. The exception to this in some designs is the joint between ceiling joist and rafter,

TRADA Standard roof truss sheets

Design sheet number	Tile weight (kg/m square)	Pitch (°)	Span limits (mm)
500	68.4	40	5000–6000
501	68.4	40	6000–7000
502	68.4	40	7000–8000
503	68.4	40	8000–9000
508	44	35	5000–6000
509	44	35	6000–7000
510	44	35	7000–8000
511	44	35	8000–9000
515	58	22–30	5000–6000
516	58	22–30	6000–7000
517	58	22–30	7000–8000
518	58	22–30	8000–9000
519	58	22–30	9000–10000
520	58	22–30	10000–11000
521	80	22.5–35	5000–6100
522	80	22.5–35	6100–7000
523	80	22.5–35	7100–8000
524	80	22.5–35	8100–9000
525	80	22.5–35	9100–10000
526	80	22.5–35	10100–11000
527	73	27.5–40	5000–6000
528	73	27.5–40	6000–7000
529	73	27.5–40	7000–8000
530	73	27.5–40	8000–9000
531	73	27.5–40	9000–10000
532	73	27.5–40	10000–11000
533	100	15–25	6000–7000
534	100	15–25	7000–8000
535	100	15–25	8000–9000
536	100	15–25	9000–10000
537	100	15–25	10000–11000
538	100	15–25	11000–12000

Fig. 4.3

Fig. 4.4

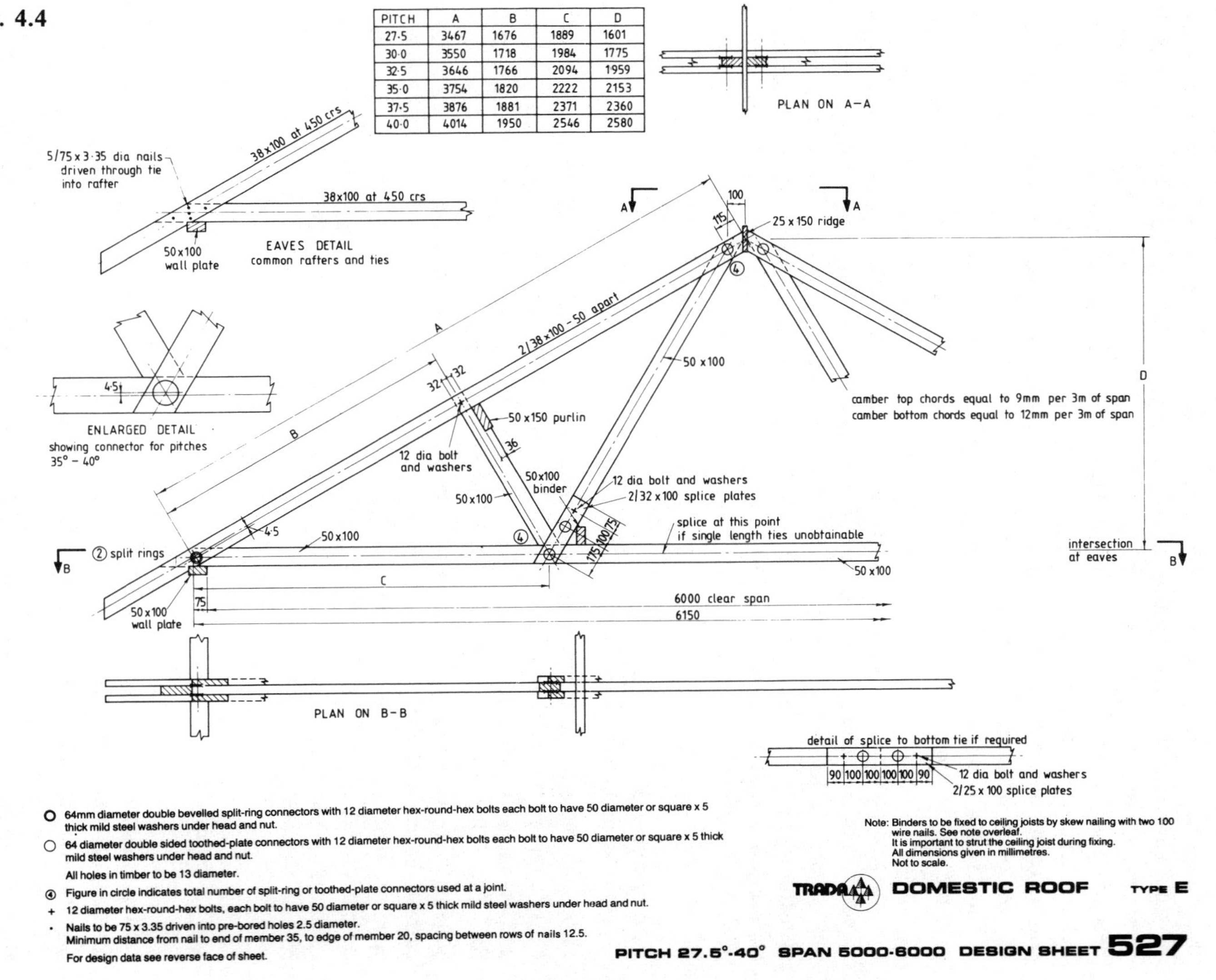

PITCH	A	B	C	D
27·5	3467	1676	1889	1601
30·0	3550	1718	1984	1775
32·5	3646	1766	2094	1959
35·0	3754	1820	2222	2153
37·5	3876	1881	2371	2360
40·0	4014	1950	2546	2580

- ○ 64mm diameter double bevelled split-ring connectors with 12 diameter hex-round-hex bolts each bolt to have 50 diameter or square x 5 thick mild steel washers under head and nut.
- ○ 64 diameter double sided toothed-plate connectors with 12 diameter hex-round-hex bolts each bolt to have 50 diameter or square x 5 thick mild steel washers under head and nut.

All holes in timber to be 13 diameter.

- ④ Figure in circle indicates total number of split-ring or toothed-plate connectors used at a joint.
- \+ 12 diameter hex-round-hex bolts, each bolt to have 50 diameter or square x 5 thick mild steel washers under head and nut.
- · Nails to be 75 x 3.35 driven into pre-bored holes 2.5 diameter. Minimum distance from nail to end of member 35, to edge of member 20, spacing between rows of nails 12.5.

For design data see reverse face of sheet.

Note: Binders to be fixed to ceiling joists by skew nailing with two 100 wire nails. See note overleaf.
It is important to strut the ceiling joist during fixing.
All dimensions given in millimetres.
Not to scale.

TRADA **DOMESTIC ROOF** TYPE E

PITCH 27.5°-40° SPAN 5000-6000 DESIGN SHEET 527

and on all designs where the ceiling tie cannot be obtained in one length, also the splice joint between the two lengths, which again uses a combination of connectors and nails. An alternative for the ceiling joist allows for it to be joined over a plate on a partition; this is not to be recommended where the construction process requires non-load bearing partitions to be installed after the roof is constructed. The roof should therefore be constructed wherever possible as an independent clear spanning structure between the two wall plates of the external walls. To avoid trusses inadvertently bearing on internal non-load bearing partitions, it is good practice to have the roof tiled and therefore under normal working load conditions to allow any deflection in the truss to take place before the partitions are fitted.

The trusses are not designed to carry water storage tank loads and thus, wherever possible, these should be directly supported from partitions below. If this is not practical then the advice of TRADA or the truss designer must be sought in order that the truss spacing may be reduced to carry the additional load. One final point on the standard roof assembly is that some difficulty may be experienced in nailing the ceiling joist tightly to its binder. A more effective connection is to use one of the readily available light galvanised metal cleats. These give a much more positive and stronger connection than traditional skew nailing. Fig. 4.5 illustrates the roof construction described above.

BOLTED TRUSS HIPS

Hips can be formed using bolt and connector trussed structures, the hip usually being supported on a 'half' truss, itself supported on one of the principals at hip peak and on the end wall plate at the hip eaves. A typical hip construction is illustrated in fig. 4.6. The remainder of the hip timber members follow basically traditional construction techniques, described in Chapter 3. The hip rafter, however, generally carries the ends of the purlins and also supports the hanger, which in turn helps to span the binder from first principal to the end wall plate. It is essential that the ceiling tie member of the half truss used in the hip end is connected properly into the binder running on the centre line of the roof to ensure an adequate tie from one end of the roof to the other.

VALLEYS

Valley construction follows the design set out in fig. 3.10, a principal truss being located on the ridge line of the intersecting roof in place of the double rafter illustrated. With equal span and pitch intersecting roofs, purlin lines will coincide and they can be fixed together with a steel connector as indicated in fig. 3.11. Unequal spans will mean a hanger from the higher to lower purlins and/or additional support again as indicated in fig. 3.11. Support may also be required for the main roof principal truss if there is no wall on the normal plate line. It is not generally possible to

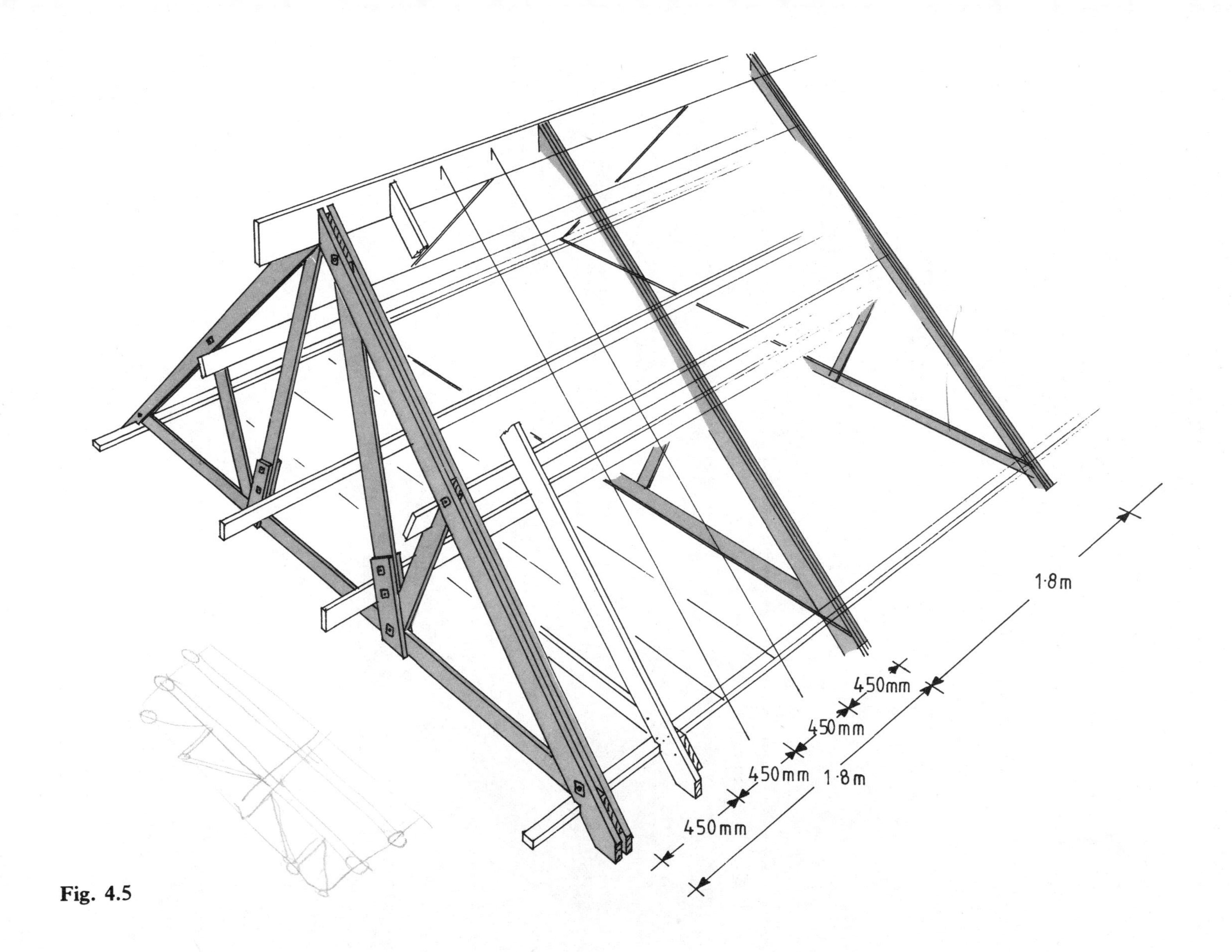

Fig. 4.5

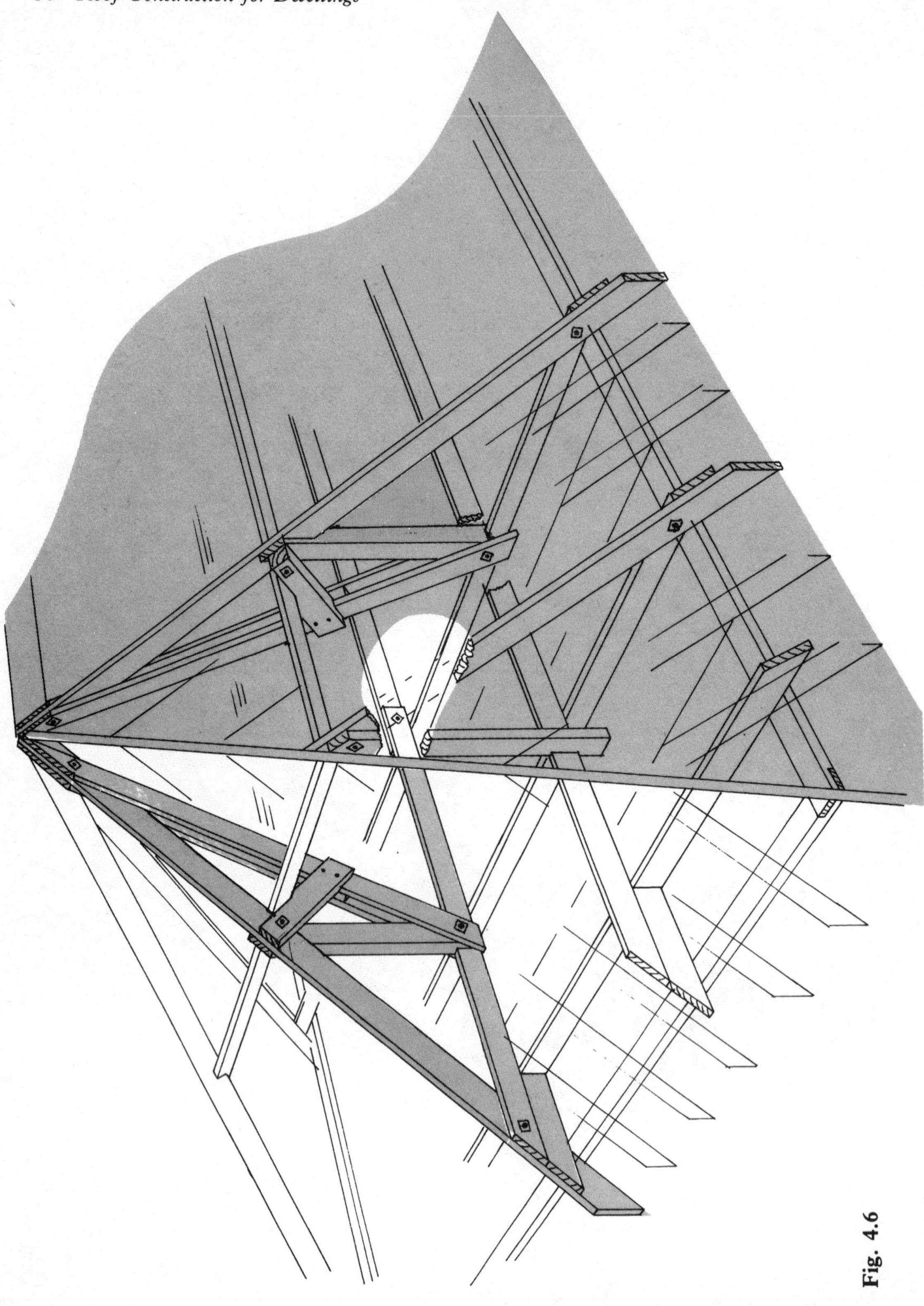

Fig. 4.6

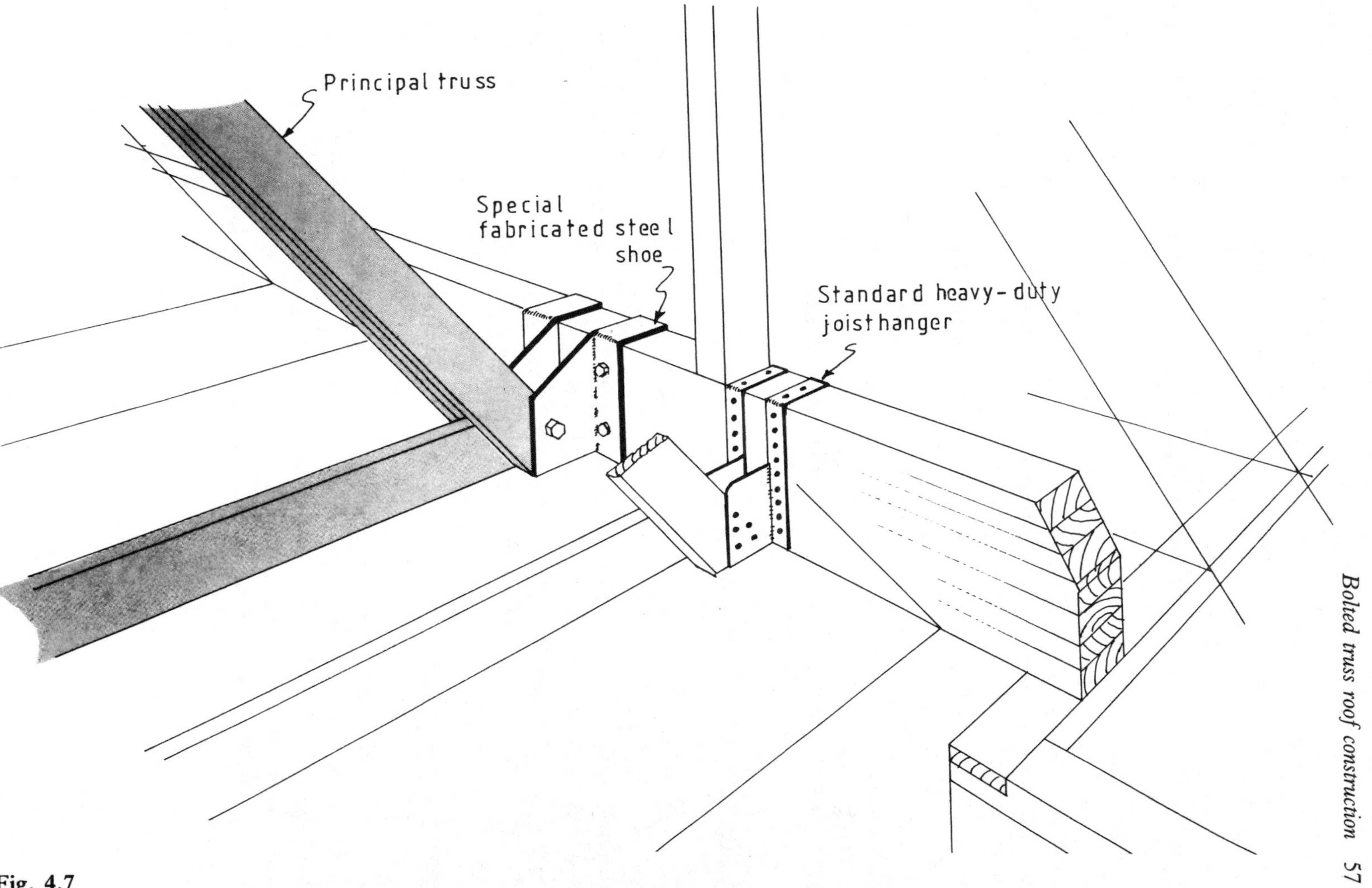
Principal truss
Special
fabricated steel
shoe
Standard heavy-duty
joisthanger

Fig. 4.7

support this principal from the first principal truss of the intersecting roof unless it has been specifically designed to carry the additional load. It is more likely that a steel or timber beam must be employed within the roof space, using a special steel shoe designed to carry the end of the principal truss. Fig. 4.7 illustrates this junction. A beam beneath the truss is a more easily constructed detail, but its depth may restrict headroom below.

Care must be taken with this steel shoe design and the positioning of the supporting beam, to allow adequate 'end-distance' for the rafter to ceiling joist connector discussed earlier. Also the positioning of the truss to shoe locating bolt should ideally be identical to the bolts used to assemble that particular joint. Assembly on site would then require temporary removal of one of the joint retaining bolts whilst the truss is temporarily supported. A slightly longer bolt is passed through the steel shoe and the truss to complete the joint.

Lighter weight steel standard hangers can be used to support rafter and ceiling joist, although the location of the supporting beam may be such that an extended length of support in the shoe similar to that used for trussed rafter hangers may be required to give adequate bearing to the supported structure members.

STRUCTURAL OPENINGS

Dormer winders are unlikely to occur in a bolt and connector roof unless a specific attic design has been produced, in which case all openings will also have been structurally designed and these must be followed. Openings for chimneys and roof hatches will occur and these can generally be dealt with as illustrated in fig. 3.18 or, in the case of small openings, as illustrated in fig. 7.26. Care must be taken to position the principal trusses at design stage to avoid such openings.

ROOF STABILITY

The bolt and connector jointed principal roof truss construction, like most other forms, should not rely for its lateral stability upon the gable end wall structure. For this reason diagonal bracing on the undersides of the rafters should be provided and of course some temporary bracing will be required at construction stage to maintain the heavy principal trusses in position. Roof bracing is dealt with in more detail in chapter seven.

CHAPTER 5

The construction of trussed rafter roofs

The majority of roofs constructed for domestic dwellings in the United Kingdom now use the punched metal plated trussed rafter construction. Over two million units are produced each year. The majority of the trussed rafters are produced in factories where capacity to both design and produce varies enormously – from those manufacturers only able to design from manuals provided by their plate supplier, to those fully competent timber engineering companies employing their own structural designers. Production capacity ranges from two hundred to three thousand trussed rafters per week, with quality also varying both in the final product and in the service offered to the customer. The punched metal nail plate connector can only be fixed using the specialist equipment needed for pressing the plates into both sides of the timber joint. It is *not* possible to fix them on site. Fig. 5.1a illustrates a typical punched metal plate joint.

There are two alternative systems to the punched metal plate fastener, one being a metal plate punched with holes which is then fixed to the timber joint with special twisted nails (fig. 5.1b). The other is to use plywood gussetted joints, the plywood being fixed either by glue with nails to hold it in position whilst the glue cures, or exclusively by nails of designed size and fixed to a specific designed pattern on the joint, fig. 5.1c illustrates such a joint. With the exception of the glued option, the latter two methods are suitable for site assembly.

PERFORMANCE IN USE

Much has been written in trade journals concerning possible problems occurring with the trussed rafter form of construction, with reference particularly to its long-term durability. An authoritative paper was prepared in 1983 by the Building Research Establishment entitled 'Trussed rafter roofs' (IP14/83) in which the results of a nationwide survey were summarized. The survey looked at the manufacture, site use, performance in service, and plate corrosion with certain types of preservative. Whilst

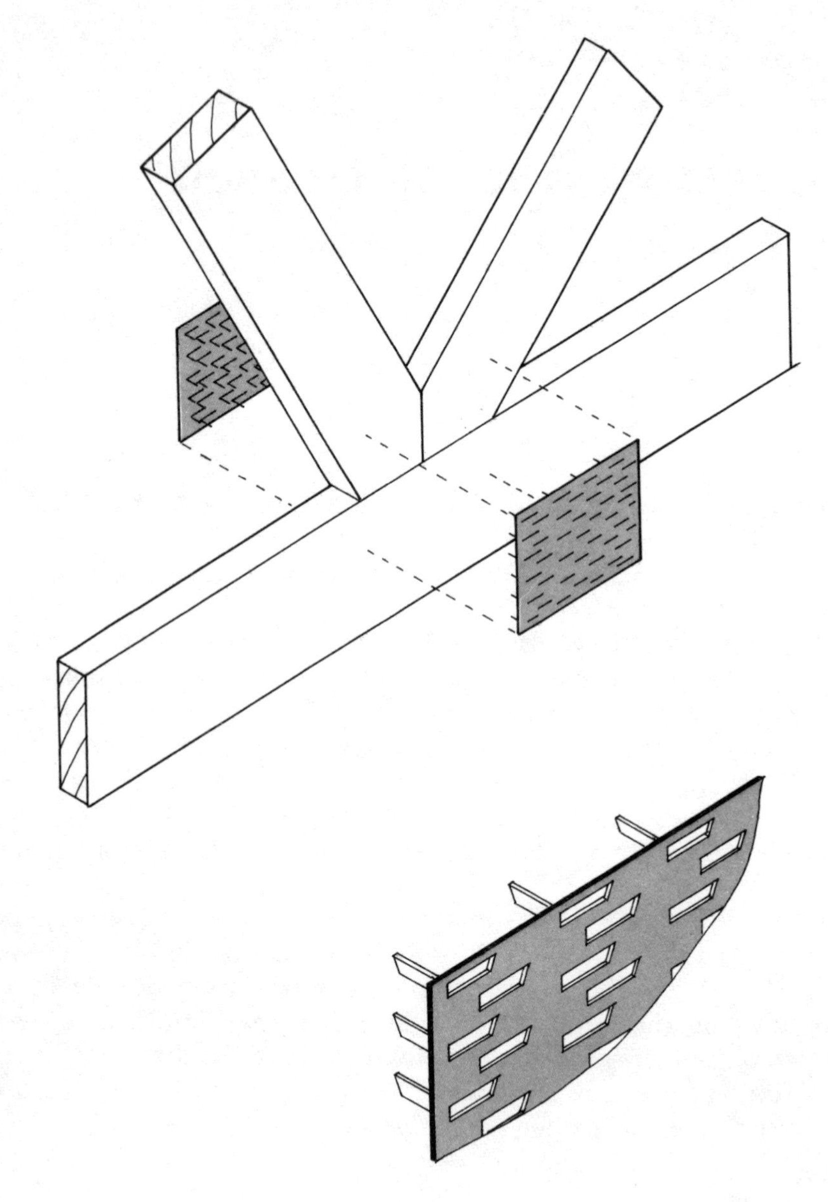

Fig. 5.1a

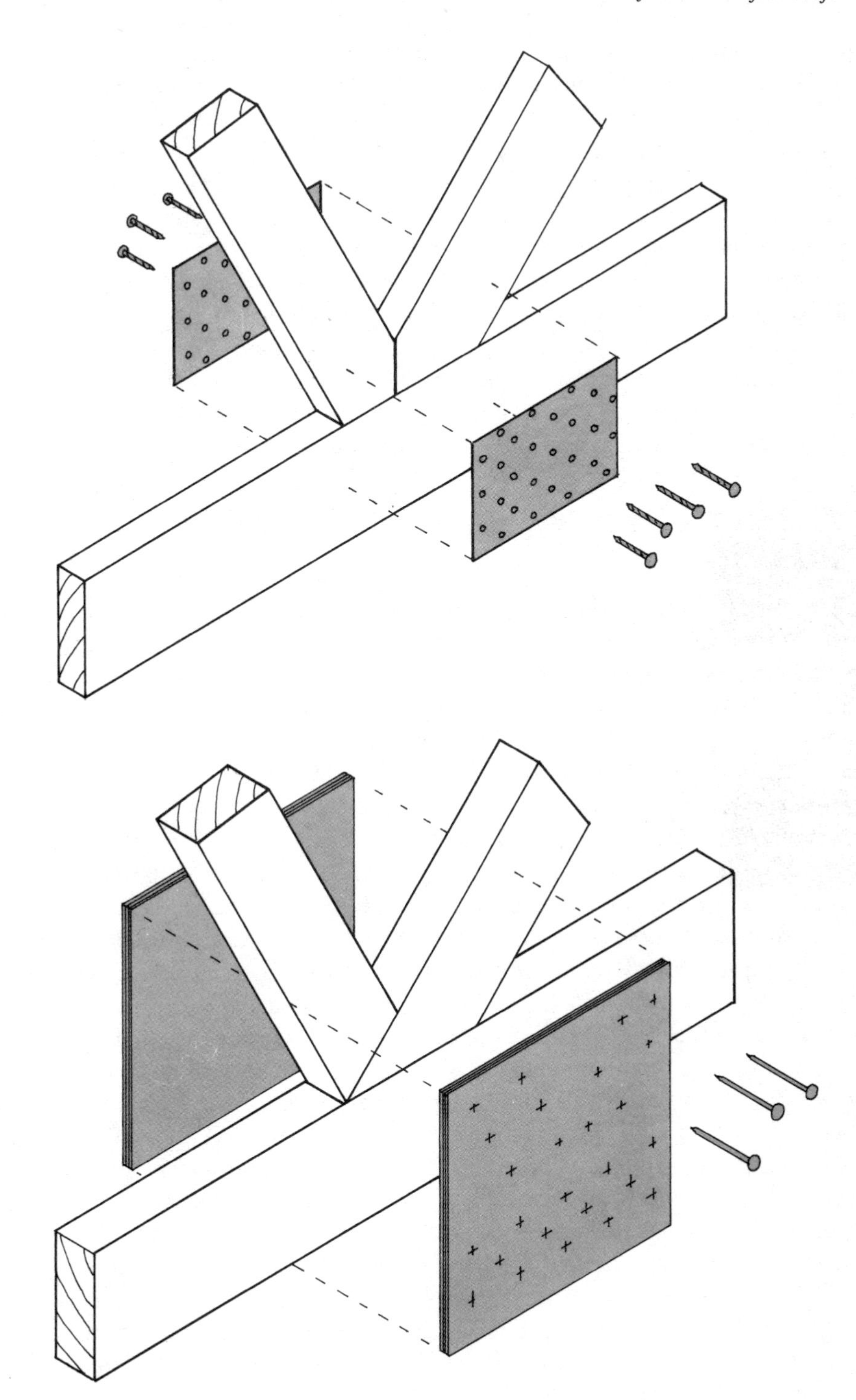

Fig. 5.1b

Fig. 5.1c

some shortcomings were found in the manufacture, the areas causing greatest concern were site handling and construction of the roof structure using the trussed rafters on site. Inspection of the trussed rafter components for compliance with the relevant British Standards before assembly into the roof structure should be made, thus overcoming the possibility of faulty structural components being installed. The problem of inadequate design and assembly information for the roof structure as a whole remains, and whilst British Standard 5268 part 3 gives some guidance for conventional gable ended roofs and the individual plate manufacturers provide standard details for hips and valleys, the builder is not generally presented with a specific set of drawings for the assembly of the roof on which he is working.

This chapter concentrates on the construction of trussed rafter roofs including hips, valleys, attics and trimmed openings. The storage and handling of trussed rafters and other timber components, and the many ancilliary details required to complete the roof are set out in chapter seven.

DESIGN

The structural design, manufacture and some aspects of construction are dealt with in British Standard 5268 part 3 1985 'Code of practice for trussed rafter roofs', and reference will be made to this important document throughout this chapter. Whilst timber sizes may be obtained from safe span tables, the design of the joint plate or gusset *must* be provided by a qualified structural engineer.

Most trussed rafter manufacturers use terminology not yet covered in previous chapters and the reader should refer to fig. 5.2 for familiarisation with terms to be used throughout this chapter. The illustration shows a 'fink' truss, the most common configuration in use today. The geometry of this configuration can be found in fig. 2.5. The standard spacing for trussed rafters is 600 mm, although 400 mm and 450 mm are not uncommon. Timber sizes are standardised throughout the United Kingdom, the timber being machined on all surfaces for accuracy in accordance with the standards set out in British Standard 4471. The timber is usually stress graded in accordance with British Standard 4978 and should be stamped with a grade mark. Whilst so called 'nominal' sizes are often quoted, i.e. 75, 100, and 125 × 38 mm, the finished section will be 72, 95, and 120 × 35 mm respectively. Any precise check on trussed rafter timber sizes must take into account moisture content. Timber of 47 mm finished thickness is frequently used for attic trussed rafters with depths going up to 220 mm for heavily loaded rafters and floor joists. Trussed rafters in excess of 11 m span must use this thicker timber, or be made of multiple trusses of minimum 35 mm thickness, permanently fixed together by the truss manufacturer at works.

An understanding of the function of the trussed rafter is essential if good roof construction is to be achieved. The trussed rafter is designed to carry only the vertical loads imposed upon it, no lateral loads are catered for. The design assumes that the trussed rafter is maintained in its truly upright position by the various bracing and restraining members. Fig. 5.2 illustrates this point. The wall plate, binders, tile battens and diagonal braces are all assumed in the structural design. All of these items

Fig. 5.2

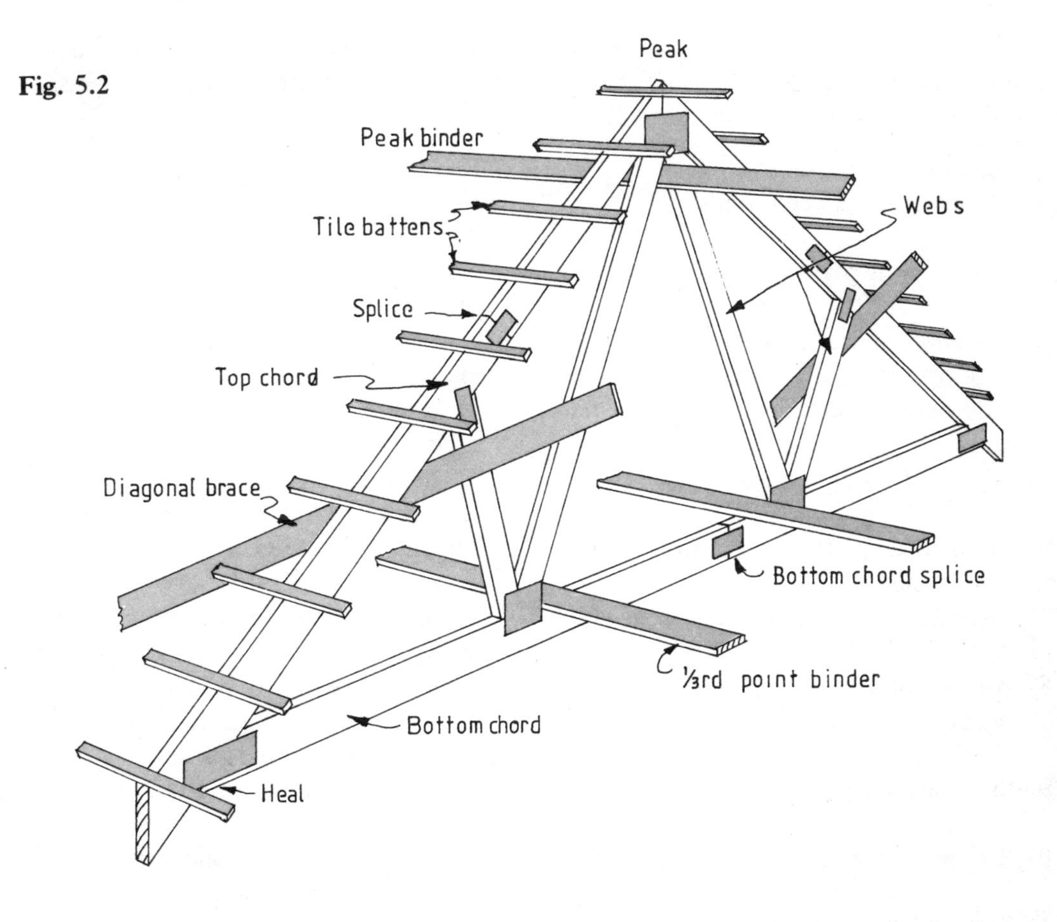

have to be site fixed and it is essential therefore that the specification for these fixings is carried out. The load on the top chord is from self weight of the truss, the tiles, felt and battens plus statutory snow loading. On the bottom chord the load is from the self weight, ceiling and insulation plus a nominal loft loading and an additional load imposed by a man walking on the bottom chord. No other loads are designed for. Wall plates must be a minimum of 75 mm in width, to avoid the load from the truss crushing the timber on the underside of the bottom chord at the point of contact with the plate. On spans above 9.3 m it is essential to use 100 mm wide plates. It is, however, common practice to use a 100 mm wide plate, this width matching precisely the width of the inner skin of a conventional cavity wall. Refer to clause 20 of BS 5268 part 3 for more detailed information.

DESIGN INFORMATION

It can be seen that, when requesting a quotation and particularly when ordering the trusses, it is essential to inform the trussed rafter manufacturer of not only the truss shape and size and number required, but also the roof tile type and ceiling specification. The best way to ensure that the design is adequate for the roof under consideration is to send a full set of drawings for the building project such that this, or an additional specification, includes the information set out below.

	Information required	Reason for data requirement
(1)	Dimension between walls	To obtain accurate span over wall plates
(2)	Size of wall plate	To obtain accurate span over wall plates
(3)	Location of wall plates or other supports for the truss	To obtain accurate span over wall plates
(4)	The overall thickness of the cavity wall	To determine the top chord overhang required
(5)	Width of soffit required	To determine the top chord overhang required
(6)	Pitch in degrees, left hand side of roof slope	To establish the geometry of the roof truss
(7)	Pitch on right hand side of roof slope	To establish the geometry of the roof truss
(8)	Any minimum top chord size requirement	Structural design may produce a rafter of smaller section if no limit placed upon the designer
(9)	Size and location of water tank (or tanks)	For design load calculations
(10)	Size and location of loft access hatch	To design any additional trusses or trimmings for openings
(11)	Size and location of any chimney stacks	To design any additional trusses or trimmings for openings
(12)	Roof tile or other covering specification giving manufacturer's name and type or a precise weight	For loading calculations
(13)	Ceiling and insulation specification	For loading calculations
(14)	Timber preservation specification requirement	Effective timber preservation must be carried out after the components are cut and before assembly
(15)	Overall length of the building	To allow the correct number of roof trusses to be calculated
(16)	Truss spacing required	To allow the correct number of roof trusses to be calculated
(17)	Site address and location of building	To inform designer of wind loads on building and to aid delivery of the goods to site

The trussed rafter designer and/or supplier should provide his customer with certain information to enable the user to check the trussed rafter construction and provide him with information to assemble the roof structure on site. For further information please refer to section ten (information required) in BS 5268 part 3.

QUALITY CONTROL

Before proceeding with the construction of a trussed rafter roof, it would be helpful to understand the quality control imposed upon the manufacturer of trussed rafters. British Standard 5268 part 3 section six 'Fabrication', sets the standards for trussed rafter production, and this standard is incorporated in the Building Regulations 1985. All trussed rafters used for dwellings should be manufactured to this standard, whether fabricated using punched metal plate fasteners, nailable metal plates or plywood gusset joints. The illustrations in figs 5.3a-j attempt a graphical interpretation of part of section six, but the reader is directed to the British Standard text itself for full information.

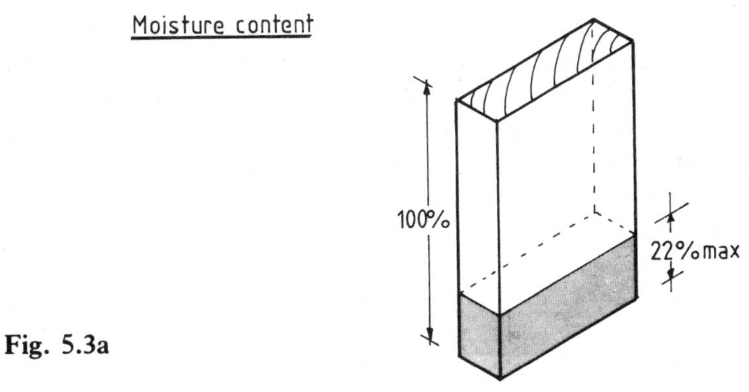

Fig. 5.3a

Fig. 5.3a shows that moisture content of the timber used in fabrication should not exceed 22%.

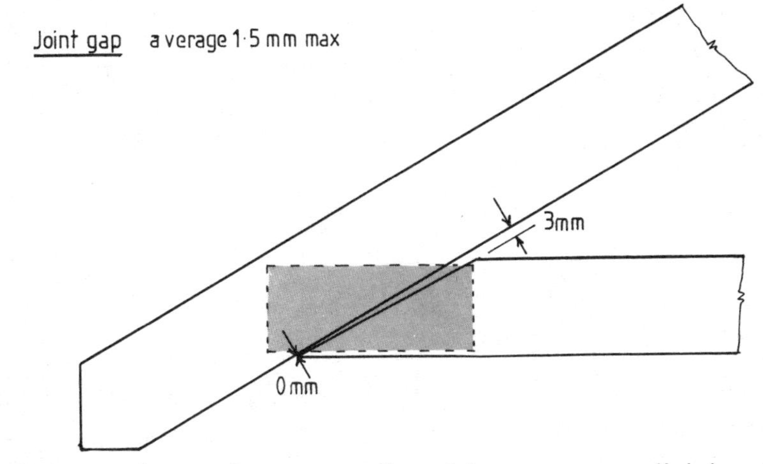

Fig. 5.3b

Fig. 5.3b covers the maximum gap allowed between two adjoining members under the punched metal plates. The average gap width should not exceed 1.5 mm unless specifically allowed for in the design.

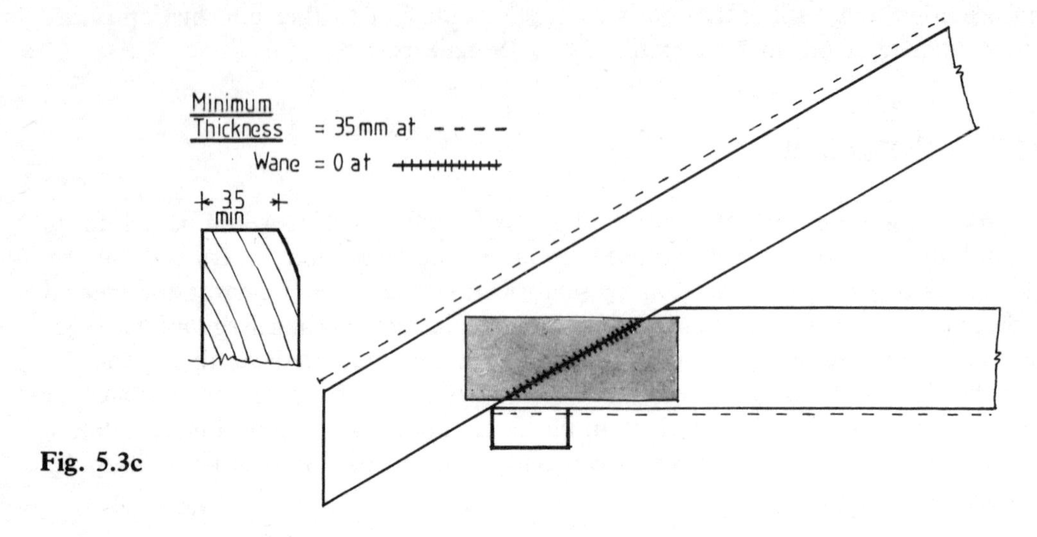

Fig. 5.3c

Fig. 5.3c illustrates that wane, the term used to describe the occasionally occurring rounded corners of the timber caused by the timber being cut near to the outside of the log of the tree, is acceptable only at certain places on the trussed rafter. It is limited on the surfaces of the trussed rafter to which other elements of the building are attached, namely the top of the top chord and the underside of the bottom chord, and within the plate area of course no wane is tolerable.

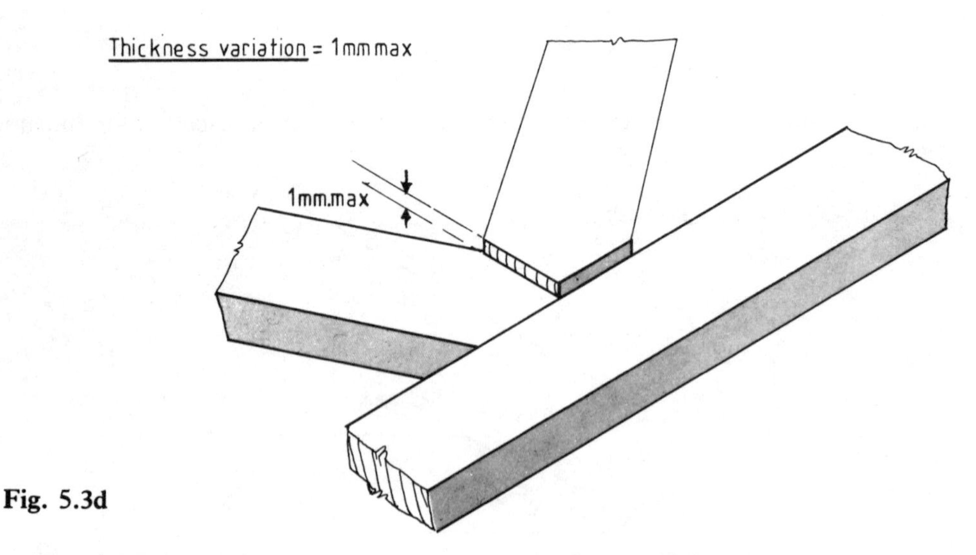

Fig. 5.3d

Fig. 5.3d shows that to ensure correct embedment of the plate teeth the difference in thickness between the members at a node point must not exceed 1 mm.

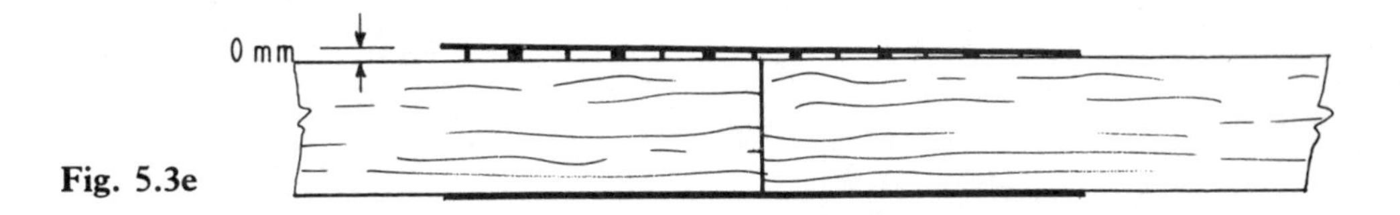

Fig. 5.3e

Fig. 5.3e ensures that metal plates are fully bedded, no gaps permitted.

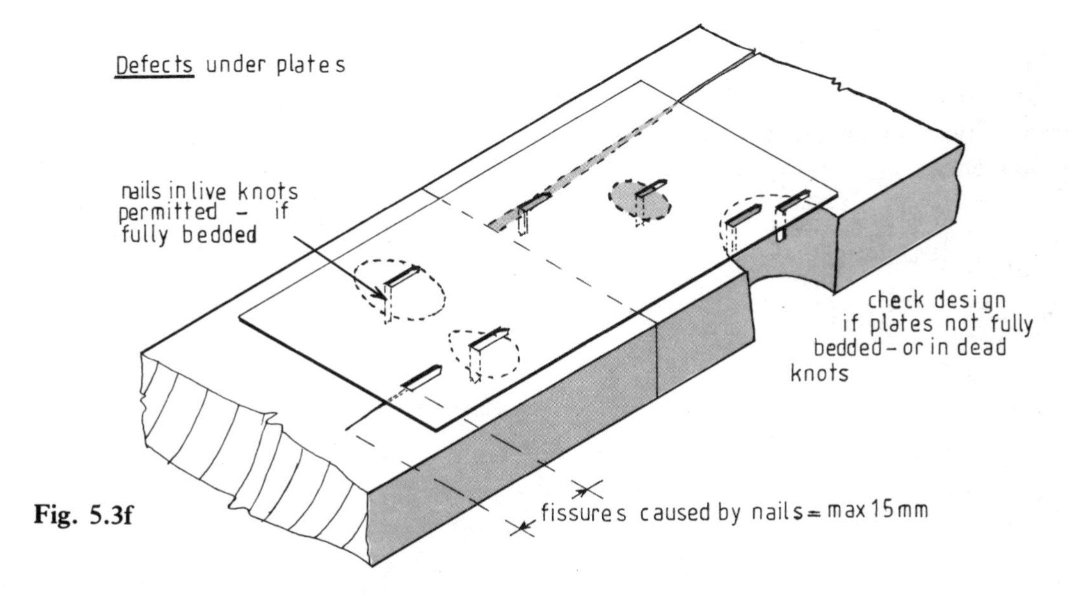

Fig. 5.3f

Fig. 5.3f illustrates that timber is a living material and not man-made, and contains features which can detract from its overall strength. Such features are splits and fissures, live knots and dead knots and, in the latter case, possibly knot holes where the dead knot has fallen out of the timber section.

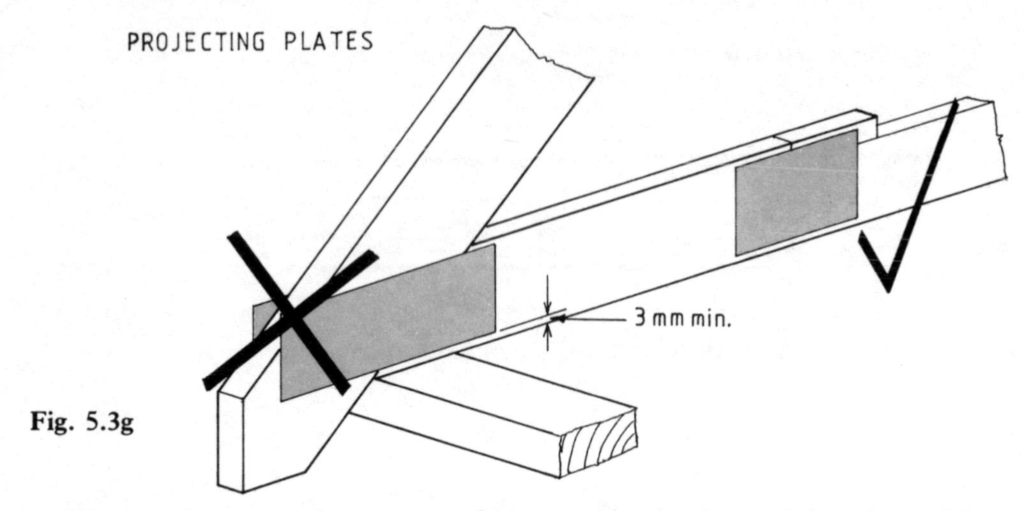

Fig. 5.3g

Fig. 5.3g illustrates that corners of plates projecting beyond the edges of the trussed rafter timbers are not allowed. Where it is impossible to avoid a projecting edge, then a timber block must be placed between the plates to protect those handling the components.

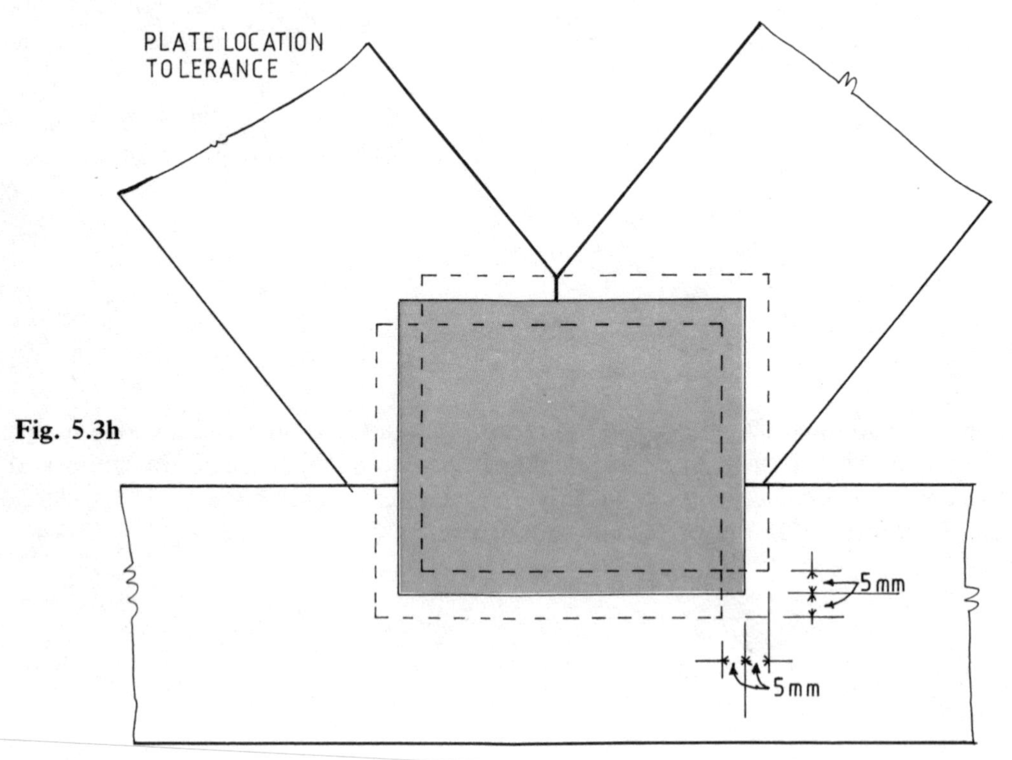

Fig. 5.3h

Fig. 5.3h illustrates plate location. Whilst this is not dealt with in section six of British Standards, there is in the design a 'plating tolerance' taking into account the possible misplacement of the plate on the joint area, whilst still maintaining the necessary safety factors in design.

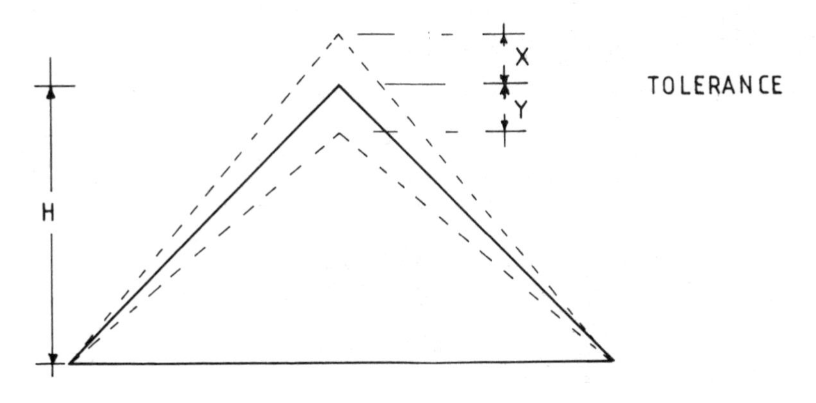

H and S for S=7·5 m. X and Y = 6mm. or less
" " S=7·5 m.+ to 12 m. X and Y = 9 mm. or less
" " S=12 m.+ " " = 12mm. "

In any continuous roof X+Y must not exceed 10mm.

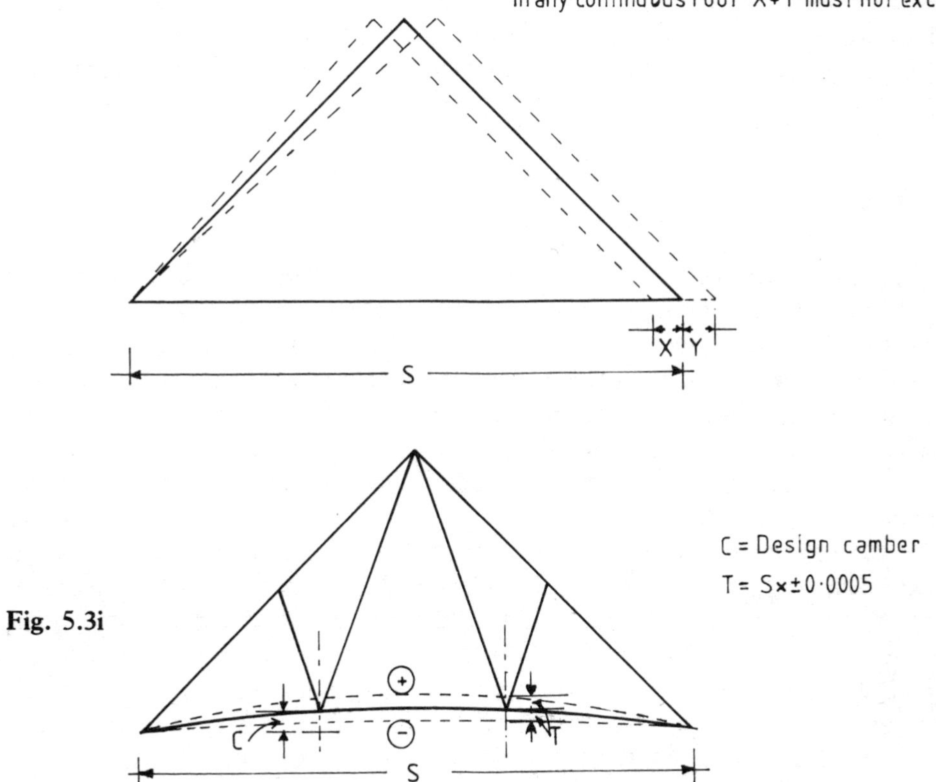

Fig. 5.3i

Fig. 5.3i illustrates dimensional tolerance. Most trussed rafters are manufactured in metal jigs, but because there is such a variety of trussed rafter spans, pitches and shapes, these jigs have to be quickly and easily adjustable. For this reason and bearing in mind the inherent natural movement of timber, it is essential that some tolerance from the design shape be allowed in production.

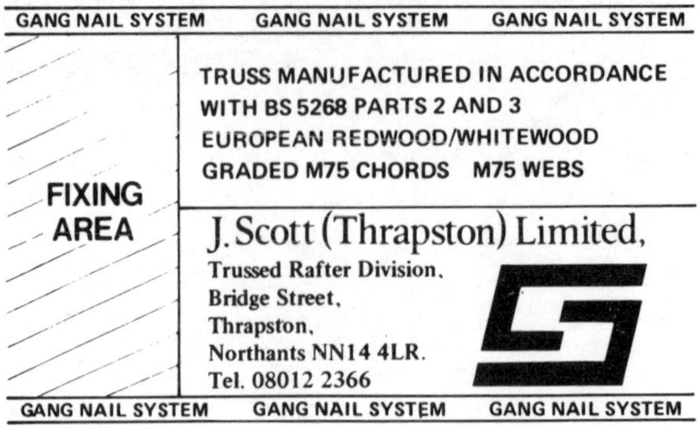

GANG NAIL SYSTEM GANG NAIL SYSTEM GANG NAIL SYSTEM

FIXING AREA

TRUSS MANUFACTURED IN ACCORDANCE
WITH BS 5268 PARTS 2 AND 3
EUROPEAN REDWOOD/WHITEWOOD
GRADED M75 CHORDS M75 WEBS

J. Scott (Thrapston) Limited,
Trussed Rafter Division,
Bridge Street,
Thrapston,
Northants NN14 4LR.
Tel. 08012 2366

GANG NAIL SYSTEM GANG NAIL SYSTEM GANG NAIL SYSTEM

Fig. 5.3j

Fig. 5.3j illustrates a marked trussed rafter. Until the publication of British Standard 5268 part 3 it was not necessary to mark the trussed rafter with the name of the company responsible for its manufacture. This has now changed, with the requirement in clause 26 of the British Standard that every trussed rafter shall be marked to show the name of the manufacturer, the species and strength grade or strength class of the timber, and finally the fact that the truss has been designed and fabricated in accordance with the British Standard.

INSPECTION

Clearly, as has been seen above, there are many aspects of trussed rafter production requiring a check to ensure their conformity with the British Standard. It must be said that the majority of trussed rafter producers are well aware of the standards required of them and produce trusses of high quality. To provide the purchaser with an assured standard of quality, many trussed rafter manufacturers belong to the TRADA quality assurance scheme for trussed rafter production. Under this scheme TRADA inspectors check the quality of trussed rafter production before accepting the manufacturer as a member of the quality assurance scheme.

Once licensed, the manufacturer has to carry out a range of checks based on the British Standard, recording the results in a log book. The manufacturer's inspector is an employee of the company, but should not be directly involved with trussed rafter production. The log book is checked by a TRADA inspector on random visits to the manufacturer. He also makes sample checks on trusses currently being produced and on those in store, pointing out any aspects of production that need attention. The fabricator is then allowed to mark the truss with a stamp provided by TRADA which naturally contains his name but also contains the TRADA mark and the manufacturer's licence number.

THE CONSTRUCTION OF A TRUSSED RAFTER ROOF

Guidelines to the correct handling and storage of timber and timber components are to be found in chapter seven.

The erection of a simple trussed rafter roof is outlined in the ITPA (International Truss Plate Association) *Technical Bulletin* number 2. The method set out below follows those guidelines but expands the techniques and extends to deal with intersecting roofs. Amendment No 1 1988 to BS 5268 part 3 gives more detailed information on bracing domestic trussed rafter roofs. Care must be taken to check that the criteria laid down in appendix A of the British Standard are observed. The standard and amendment should be to hand when reading this section.

The roof is assembled generally with nailed joints. Galvanised nails must be used, the rough surface giving a better grip and the galvanising protecting against rust. All other steel work to be fitted into the roof will be galvanised, as are the punched metal plates on the trussed rafters.

A start should be made by studying the drawings and then by checking that all timbers, fixings and trussed rafters are to hand. Check also the spans, plates, pitch and overhang on the trussed rafters. Assuming that the wall plate is bedded and the mortar set, and that any plate straps have been fitted, the trussed rafter centres should be marked out on the plate. Reference should be made to the drawing for location of chimneys, loft access traps and other openings in the roof structure. On timber framed housing the precise setting out is usually given, in order to ensure that the trussed rafters are located directly over wall panel studs.

Check the span over the actual wall plates against the truss to be used. A note here on tolerance will be useful. An overspan of 20 mm, 10 mm each side, is a sensible tolerance on the trussed rafter, thus allowing some variation in line by the wall plate without making it necessary to notch the underside of the top chord of the truss to

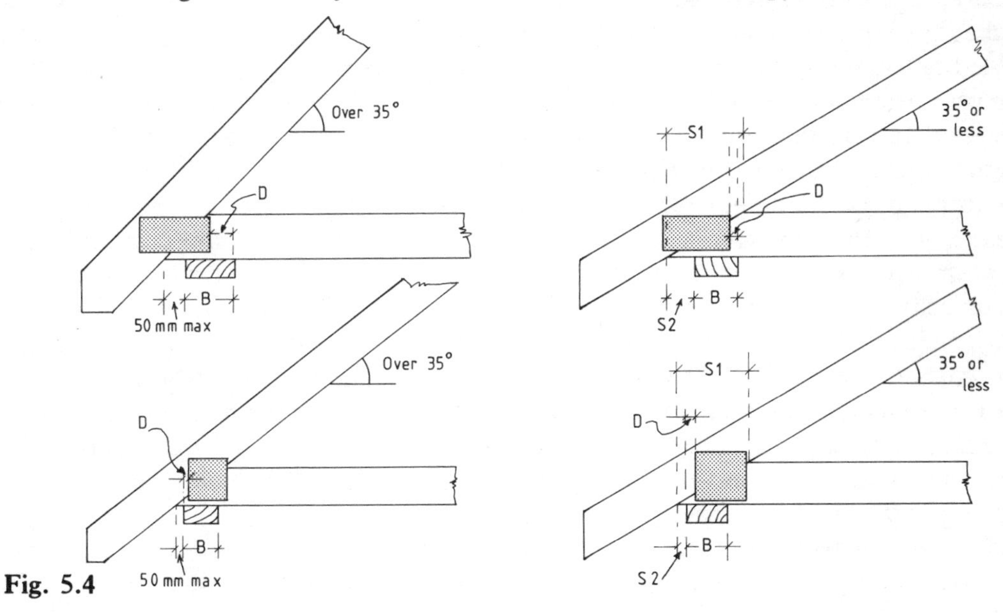

Fig. 5.4

ensure correct bearing on the bottom chord. Larger errors may occur and limits of tolerance are set by British Standard 5268 part 3 clause 42K. In fig. 5.4, for pitches 35° or less, S2 must not exceed S1/3 or 50 mm whichever is the greater; for pitches greater than 35°, S2 must not exceed 50 mm, S1/3 does not apply. On all pitches 'D' must not exceed B/2.

S1 = Length of scarf joint
S2 = Bottom chord cantilever
D = Width of wall plate not directly under connecter plate
B = Width of trussed rafter support

Under no circumstances may a trussed rafter be cut or altered in any way on site without a design drawing and specification provided by the trussed rafter designer.

The practice of 'skew' or 'tosh' nailing the trussed rafter to the plate is not recommended, as this can easily damage both the heel joint timber and nail plates. Truss clips made from galvanised steel are available from most truss suppliers and builders merchants and these should be fixed to the wall plate at the trussed rafter setting out marks. The truss clip is illustrated in fig. 7.10. Under certain design conditions truss straps may also be required, but these could be fitted just prior to roof covering.

The object at this stage of construction is to erect a portion of the roof that will be stable, from which the remainder of the roof can be constructed. Diagonal braces should be used for stabilising the roof, *not the gable end* even if at this stage it has been built. The method now described will also work for the centre section of a hip roof. Reference should be made to fig. 5.5.

The first trussed rafter, A, should be positioned a few trusses away from the gable or hip peak, such that its peak coincides with the top diagonal brace F. Holding this truss truly vertical, temporary diagonal brace B should be fitted on both sides of the roof. The brace should be well nailed to the trussed rafter as close to the web joint as possible but not in it, and to the wall plate. Temporary fixings should be made using double-headed shuttering nails to avoid damaging timbers when they are moved.

Prepare a temporary batten for both sides of the roof, long enough to reach from trussed rafter A to the gable. The battens should be marked with the truss centres by reference to the wall plate marks and shuttering nails driven in until the points just show through. Before standing truss C in place, ensure that it is the correct way round to match truss A. This can be done by checking that the bottom chord splice plates will be in-line down the length of the roof. If no splice plate is fitted, then check truss for similarity whilst still in stack. *Do not hand the trusses,* failure to carry out this check could result in poor roof alignment caused by the peak being slightly off true centre. Stand trussed rafter C in position and fix to wall plate with truss clip. Fit the temporary battens to trussed rafters A and C on both sides of the roof, the prefixed nails now easily locate the trusses and provide a free hand to help stabilise the structure. The remainder of the trussed rafters in this first section can now be fixed, each nailed to the temporary batten on both sides of the roof.

Before proceeding further, check that the trussed rafters are truly vertical. Acceptable tolerances for plumb are set out in British Standard 5268 part 3 section 7, clause 30.1, table 11:

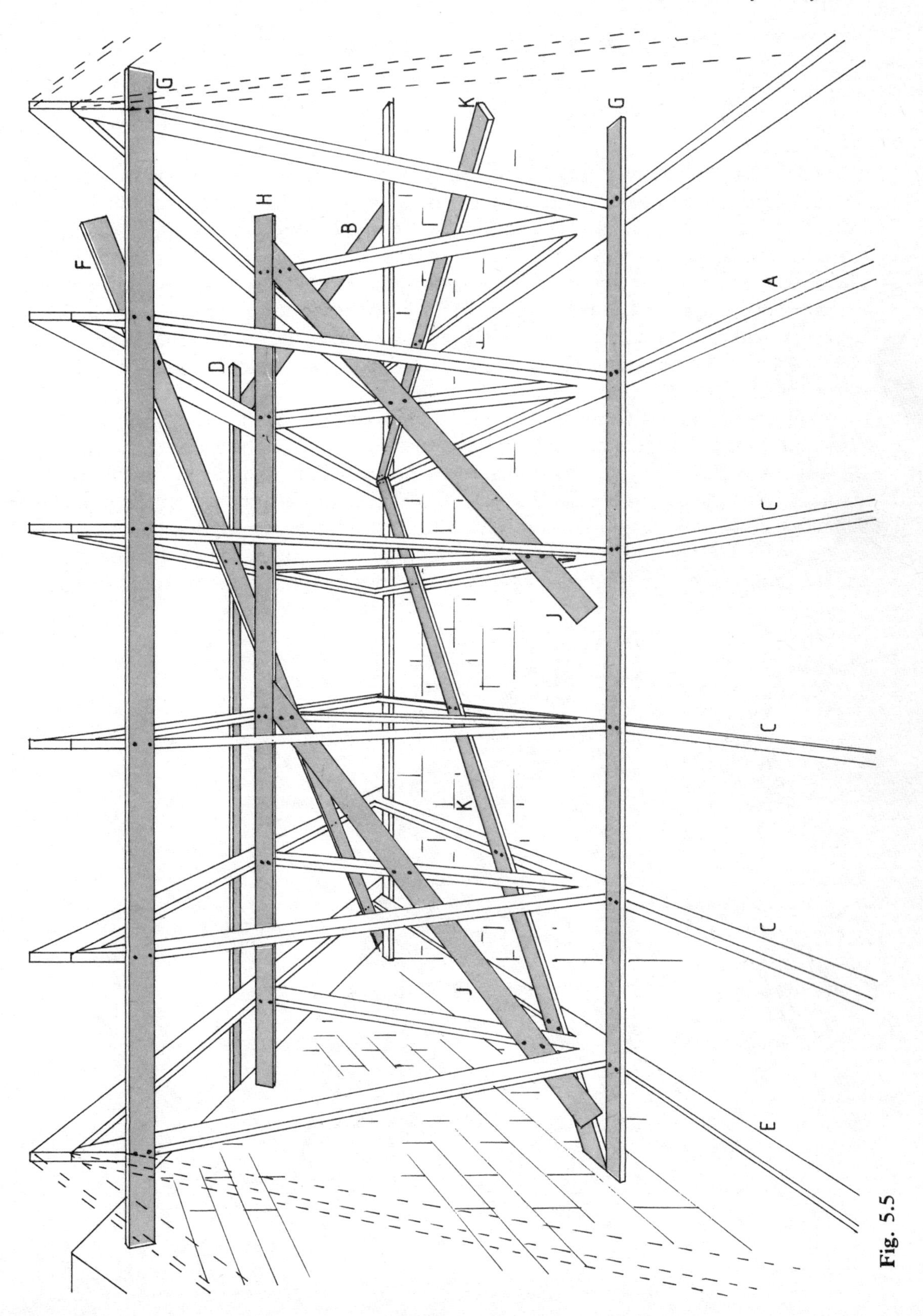

Fig. 5.5

MAXIMUM DEVIATION FROM VERTICAL

Rise of trussed rafters (m)	1	2	3	4 or more
Deviation from vertical (mm)	10	15	20	25

Diagonal brace F, 22 mm × 97 mm, should now be fitted to the underside of the top chords of the trusses on both sides of the roof at an angle of approximately 45° to the rafters. At least four diagonals 'F' should be fitted to each roof. Two 3.35 mm diameter × 65 mm long galvanised round wire nails should be used at each trussed rafter where the diagonals cross. The diagonals must extend down to the wall plate, and should be let in to the top of the plate and nailed as above. If the brace has to be lapped in its length, then the lap should cover two trussed rafters and each piece be nailed to the trussed rafters as described. Fig. 5.6 illustrates the diagonal brace lap, viewed from beneath the rafters; other binders are omitted for clarity.

Longitudinal binders G, again 22 mm × 97 mm in section, should now be fitted at peak and third point. Nailing is the same as for the diagonal, but check the centres of the trusses and the bottom chord alignment before fixing. Failure to do this could result in ceiling fixing problems – see fig. 5.7, the trusses illustrated, not being located at precisely 600 mm or 400 mm centres, will not readily accept ceiling boards which are generally 1200 mm wide. This necessitates a batten being nailed to the side of the trussed rafters to correct the spacing error. The ends of these binders should be tight

Fig. 5.6

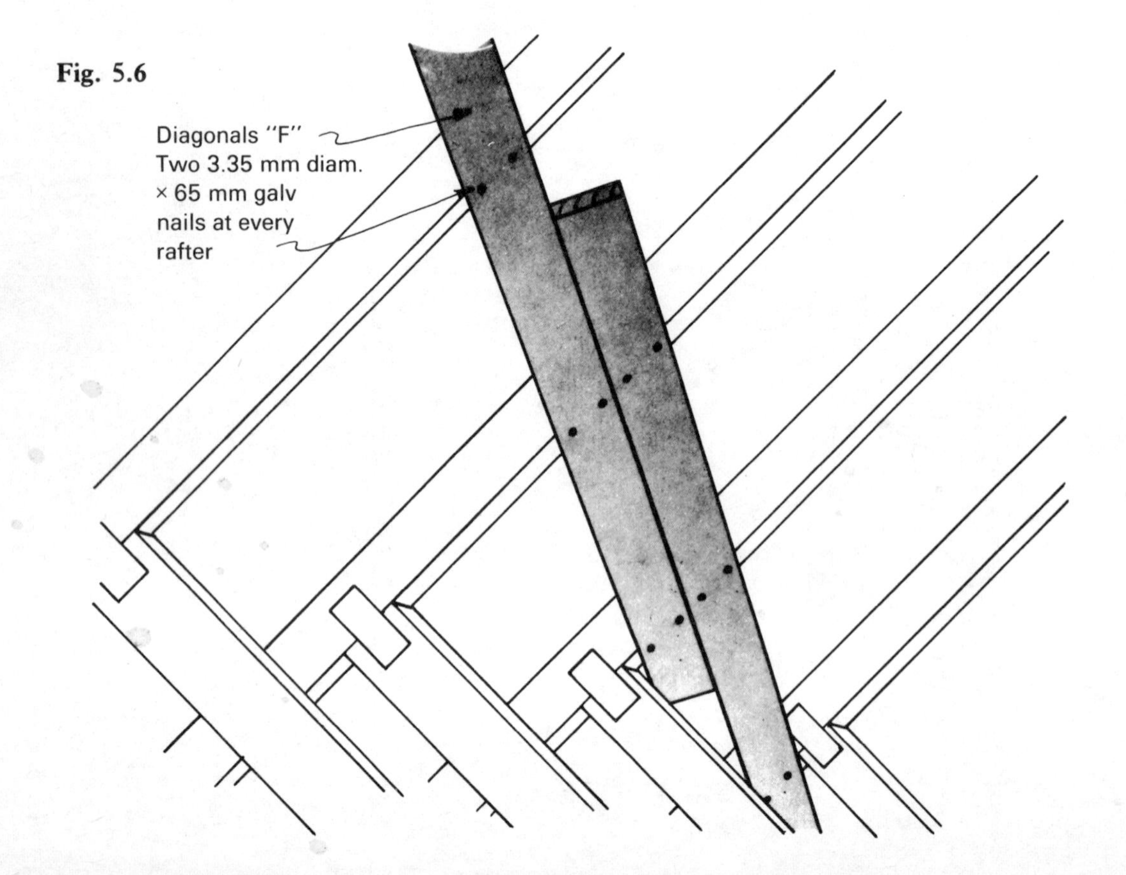

Fig. 5.7

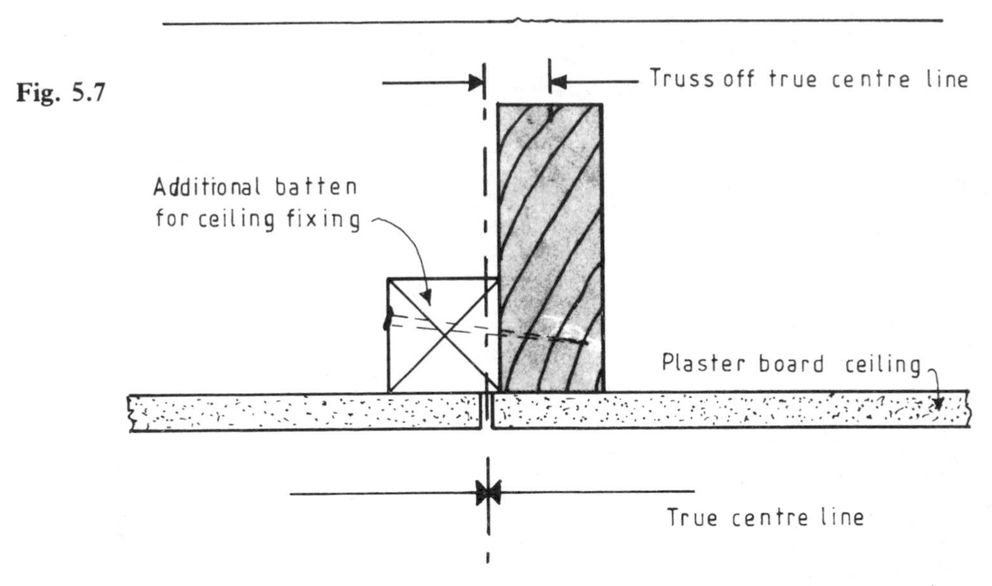

against the gable wall. On timber framed housing binders should be nailed to the gable end panel. Blocking pieces will be required, the depth of the bottom chord sufficient to enable the third point binders to be fixed to the panel frame.

On most domestic roofs the binders and bracing described thus far will be adequate to stabilise temporarily the trussed rafters themselves, at one end of the roof. The procedure is repeated at the opposite end of the roof, creating two stable elements, and finally the centre section is filled in, taking care to lap the longitudinal G over two trussed rafters.

The roof structure has to absorb the load from wind blowing on the gable ends, causing pressure at one end and suction at the other. British Standard 5268 part 3 appendix A gives standard bracing, provided certain conditions of wind speed, pitch and span are not exceeded. Specification must therefore be checked before proceeding.

Continuing with the construction, it will be assumed that the standard wind bracing is satisfactory. Further diagonal braces F must be fitted to form 'X' formations down the whole length of the building. The longitudinal binder H should be fitted, timber section and nailing as before, from one gable to the other, lapping as required. Fit this to both slopes of the roof close to the web to top chord joints. Diagonals J should now be fitted to the short webs, but only on trusses greater than 8 m span. Although not critical, the direction of slope should be from the bottom chord at the gable to the top chord, the slope will reverse at the centre line of the building. Finally, diagonal K should be fitted on top of the bottom chords, these should be placed at approximately 45° to the bottom chords and produce a 'W' formation from one gable to the other. Diagonal 'K' is not a requirement of BS 5268 part 3 amendment No 1, but in the author's opinion is advisable. Appendix A of the British Standard recognises the lack of lateral stiffness of house supporting walls exceeding 9 m long between buttressing, and restricts its bracing information accordingly. Many houses at construction stage have walls on which trusses are supported with no buttressing from internal partitions, and are no more than a series of slender columns of cavity brickwork connected only by steel or concrete lintols. At roof construction and tiling stage the roof does not have the benefit of the plasterboard ceiling diaphragm. In timber framed housing brace K is essential.

The roof is now capable of withstanding the wind loads placed upon it, but the gable walls are not connected to it. On the pressure side the wall would be blown onto the roof and would probably be secure, provided the blocking had been fixed between the wall and the first truss, and between the first and second trusses (see fig. 7.18). On the suction side, the wall would be sucked out. The gable walls must therefore be fixed securely into the roof structure with metal straps. These gable restraint straps apply to all forms of roof construction and are therefore detailed in chapter seven under a separate heading. British Standard 5268 part 3 appendix B gives an appropriate standard for trussed roofs. Details of strapping roofs to walls are also given in Approved Document A of the 1985 Building Regulations.

To complete the roof structure, gable ladders, water tank platform, barge, fascia, soffit, ventilators and eaves insulation controllers must be fitted. These items are dealt with in detail in chapter seven. In addition to the bracing A-K described above, some extra bracing may be required by the truss designer, to prevent compression buckling of long web members. This additional bracing will take the form or more rows of brace H, but fitted at the centre of the web length.

HIP END ROOFS

The construction of trussed rafter hip end roofs varies from one truss system to another. Precise construction details are therefore given in chapter six under each plate manufacturer's heading.

On roofs less than 5 m span almost traditional hip construction may be used, as detailed in chapter three. In excess of this span most systems use prefabricated hip trusses with, in some cases, only the extreme lower corners of the hip being of site cut and fitted construction. More details of the variation between trussed rafter manufacturer's hip construction is given in chapter six.

Rafter diagonal bracing is not generally required in a hip roof, the hip itself being a ridged diagonal structure. Longitudinal binders G must be continued from the main roof into the hip construction, plus additional binders H fixed flat on top of the hip trusses' top chords. Diagonal K should be installed (see fig. 5.5).

Most trussed rafter hip constructions involve a compound girder assembled from two, three or in some cases four individual trussed rafters. The nailing together of these individual components to form the girder is critical, and should be carried out by the truss manufacturer at works, if not then a nailing pattern must be provided by the roof designer. The setting out of the girder and intermediate trussed rafters is also critical and dimensions given on drawings must be carefully followed.

VALLEYS AND INTERSECTING ROOFS

The construction of valleys on trussed rafter roofs varies little between plate systems, most using a set of reducing trusses imposed upon the main roof structure. The designs are however dealt with separately under each system heading in chapter six.

Complex intersections normally have special design drawings supplied for them by the trussed rafter manufacturer. The solutions follow a pattern and typical details are

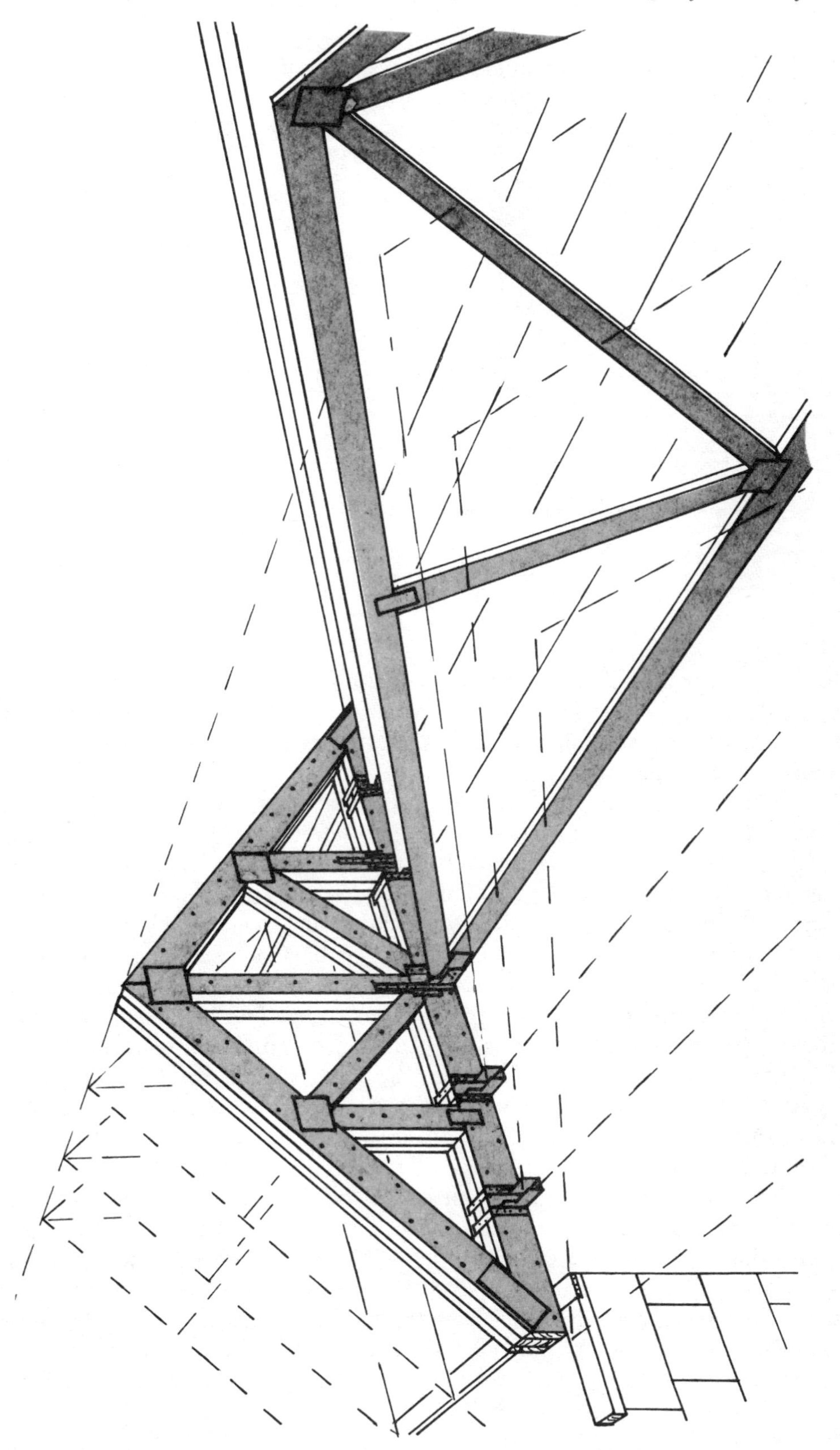

Fig. 5.8

Fig. 5.9

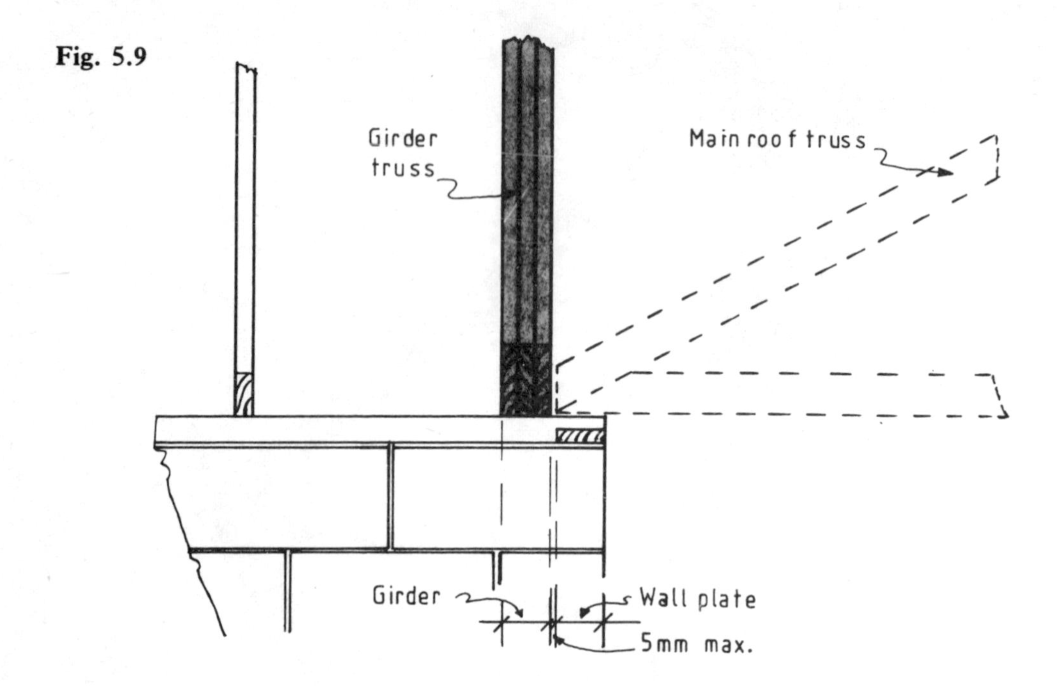

given below. Firstly, fig. 5.8 illustrates the construction of a roof intersection on which the main roof cannot be supported on a wall or beam, but is instead supported on a girder or compound truss. Secondly, in fig. 5.11, a further complication is caused by a hip occurring at the intersection.

The location of the girder truss must be such that it will carry the main roof standard truss, the overhang of the standard trussed rafters simply being removed on site if special trusses have not been provided. If the overhang is removed on site, care must be taken not to cut too near the heel joint plate, a minimum distance of 10 mm is recommended. See fig. 5.9 for the girder to standard truss detail.

Girder or compound trusses are made up of two or more specifically designed trussed rafters. The girder has to carry considerable loads from the end of the standard trussed rafters of the main roof, the load concentrated on the bottom chord. To stiffen this member, a deeper timber section is often used and a different configuration of webs dividing the bottom chord into four or more bays, such as those shown in fig. 2.8. The girder should be fully assembled in the factory, arriving on site as one unit not separate trusses. If this is not the case, then a nailing specification must be supplied by the trussed rafter manufacturer. The correct nails and spacing must be used if the individual trussed rafters in the girder are to function as one. The load from the main roof is concentrated on one side of the girder and inadequate nailing will lead to deflection giving a poor ceiling line and, more seriously, possibly a failure (see fig. 5.10).

The shoes used to support the main roof trussed rafters on the girder must be designed for the function. Normal joist hangers are not adequate for two reasons:

Fig. 5.10

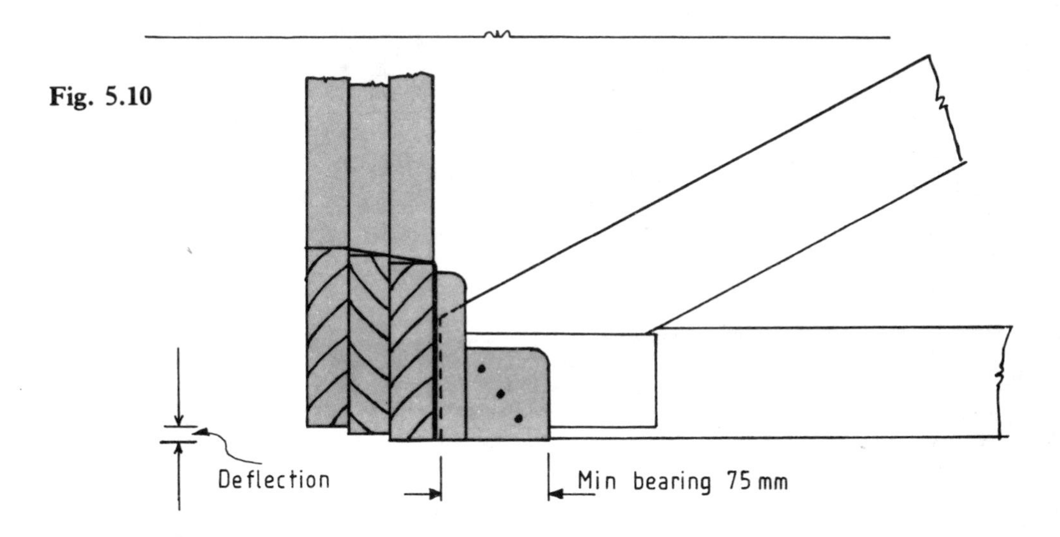

(1) They are unlikely to be strong enough
(2) They will have insufficient bearing for the truss. If a 100 mm wide wall plate is needed, then 100 mm bearing is needed in the shoe

Standard truss shoes are available from most trussed rafter manufacturers and builders merchants. The shoes must be fixed to the girder with the correct number and size of nails, screws or bolts specified.

Referring now to the hip end valley intersection illustrated in fig. 5.11, the loading on the girder is even further complicated by the large point load from the hip girder truss. A special galvanised steel shoe is required to transfer this load from the hip girder to the supporting girder. Fixing must be by bolts, screws and nails being inadequate. Bolts should not be placed in the area of the connector plates, one solution being to bolt above the junction through one of the girder webs (see fig. 5.12).

TRIMMINGS FOR OPENINGS

Small openings for flues or loft hatches are dealt with in chapter seven under the appropriate headings. Details below apply to large openings, i.e. those greater than two standard trussed rafter spacings. Such openings can be provided using specially designed girder or compound trusses either side of the opening, or by taking support from the structure itself, in which case special trusses will be required for the infill area.

On the true trimmed opening, a purlin or purlins must be fitted between the girders onto which common rafters are fixed. The ceiling joists will be supported on a purlin at bottom chord level on the girder truss. All purlins and binders must be fixed to the trussed rafters at the node or joint points. It is good practice to have the common infill

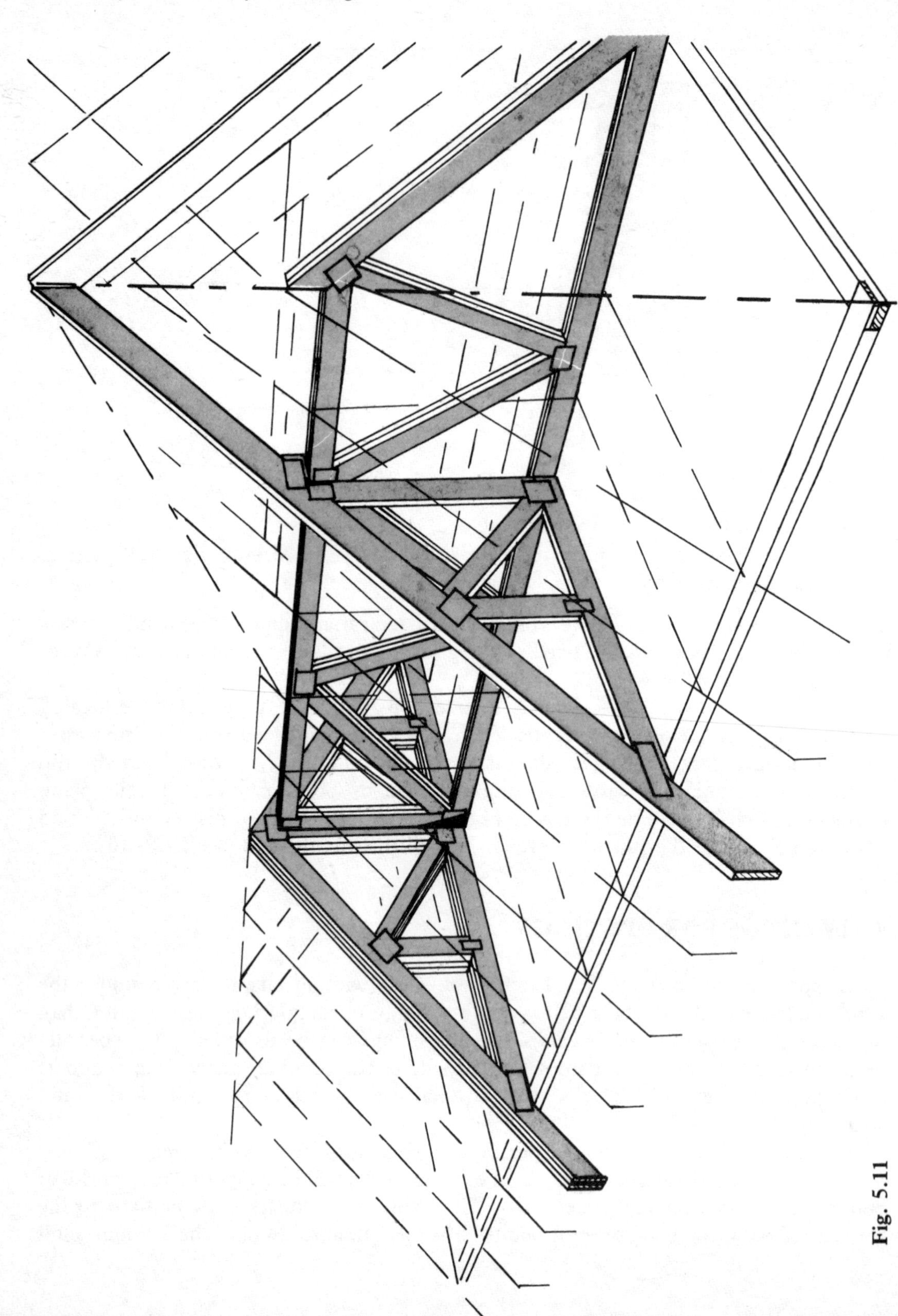

Fig. 5.11

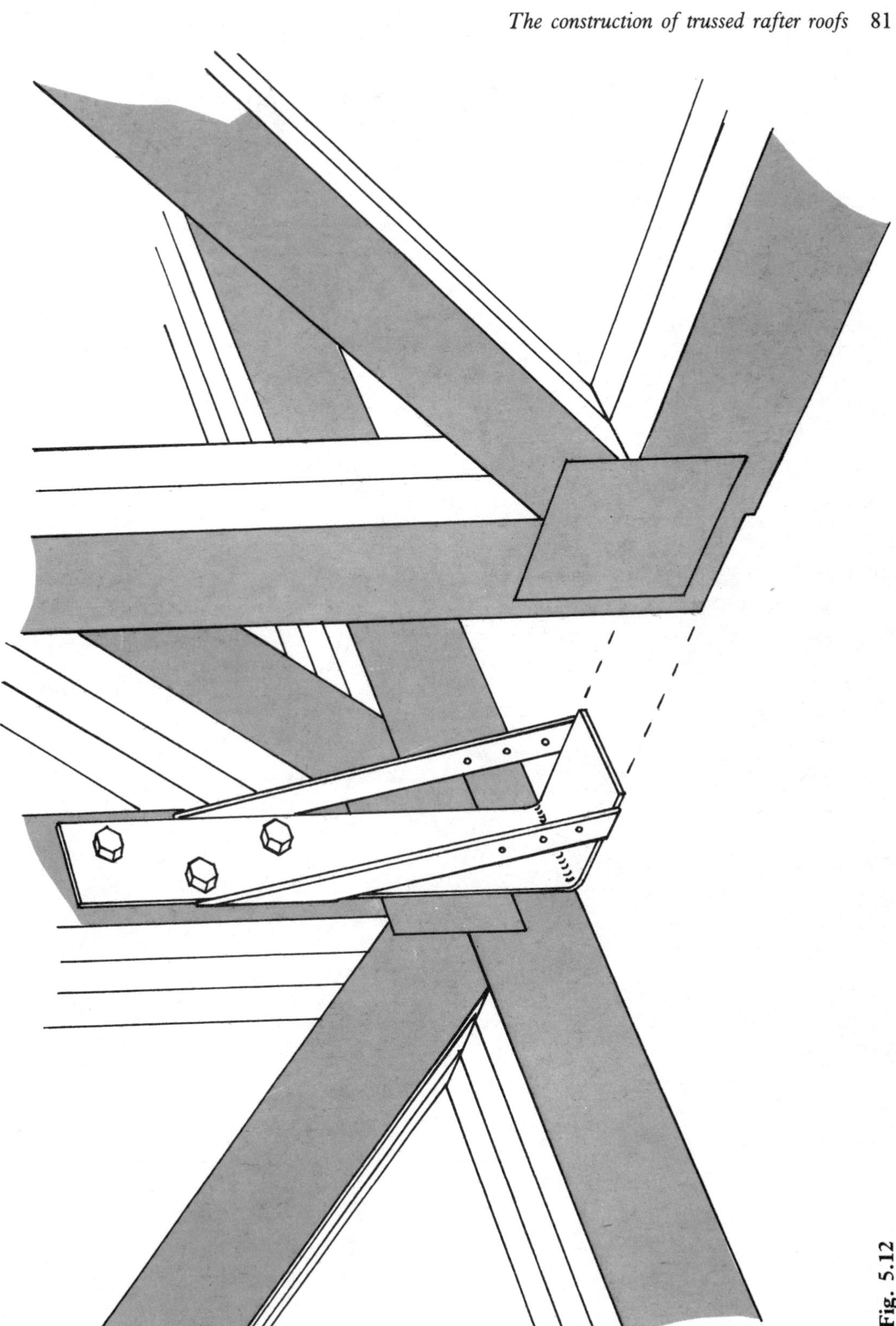

Fig. 5.12

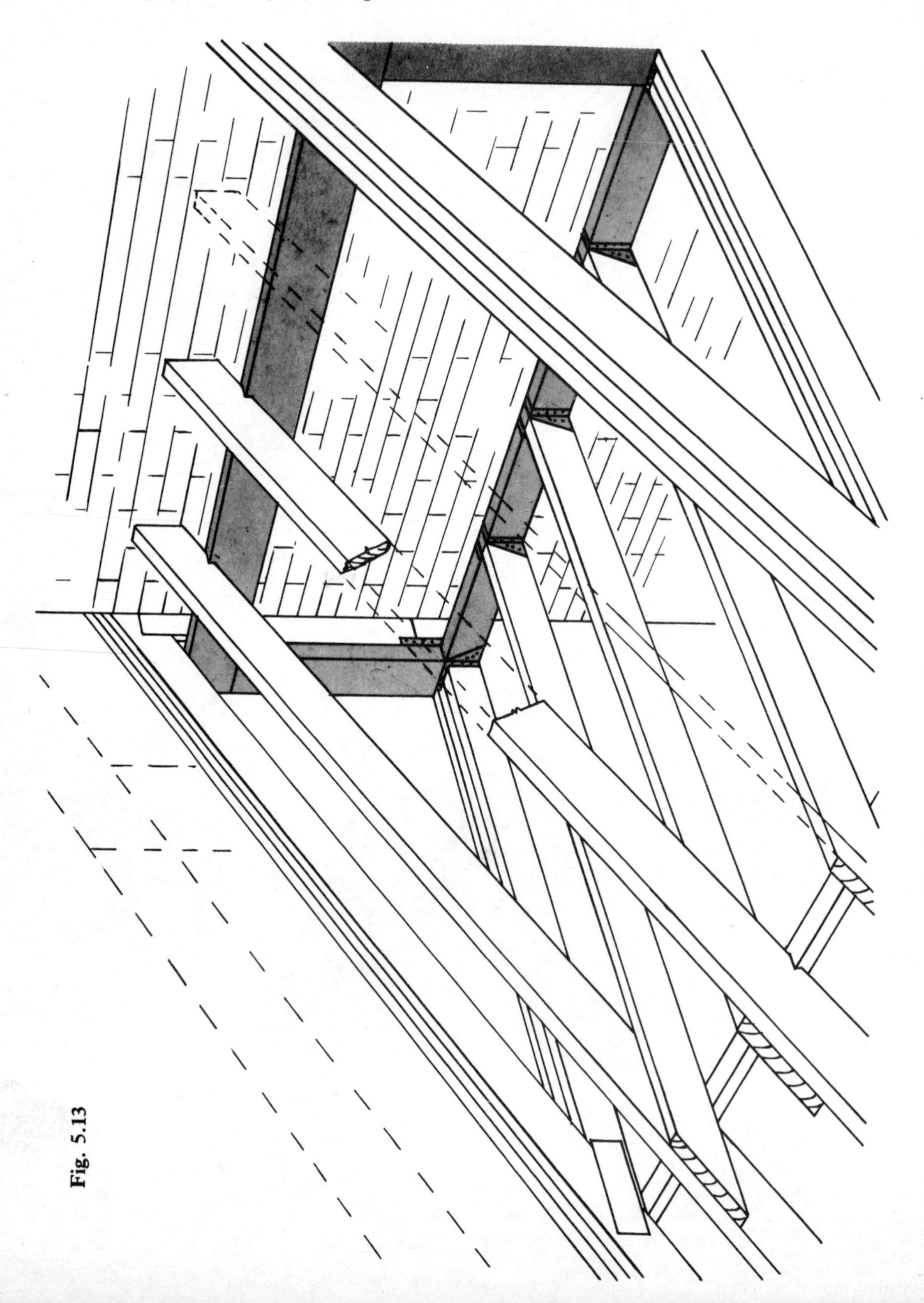

Fig. 5.13

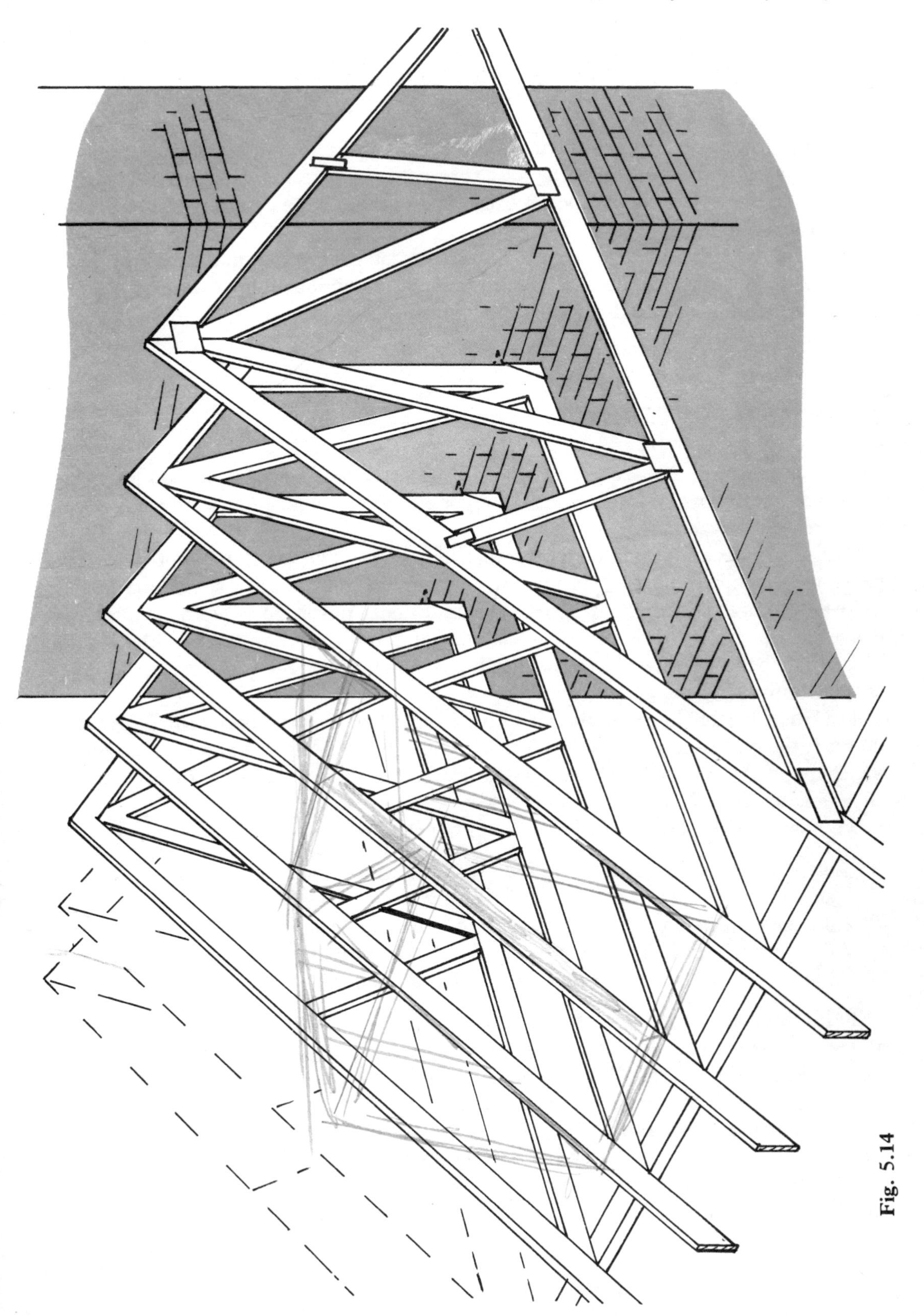

Fig. 5.14

rafters 25 mm deeper than the top chord of the truss to allow for birdsmouthing over wall plate and purlin (see fig. 5.13).

The alternative method, using the structure passing through the roof for support, is suitable for particularly large openings but does presuppose that the structure is built in advance. The special shaped trusses are shown in fig. 5.14 and can be supported either on steel shoes built into the structure or on a corbel. Care must be taken to stabilise the upper part of the special truss, this should be done with both binders and diagonal bracing.

Note that timber located adjacent to flues must conform to the rules laid down in the Building Regulations concerning the proximity of combustible materials. See the section in chapter 7 'Trimming small openings'.

ATTIC AND LOFT ROOFS

For the purposes of this section an attic will be defined as a roof void used for living, bedroom, bathroom and kitchen, playroom or studio use, whilst loft will be a roof void intended for storage only.

One of the disadvantages of the normal fink or fan trussed rafter roof is that the many web members obstruct easy access within the roof void, a space traditionally used for storage of household items. The traditional purlin and common rafter roof described in chapter three gives a relatively unobstructed loft space, although the collars may be well below a desirable headroom height. The TRADA roof described in chapter four also gives a very usable loft void.

The trussed rafter roof can provide either a loft or an attic, if specifically designed to do so. If the roof is designed with a loft space which the designer knows, or because of the shape of the internal members of the truss, can provide a room, then he should allow for domestic floor loading.

We now consider four main types of loft or attic roof:

(1) Loft void – trussed rafters carrying roof load with separate floor joists for storage load
(2) Loft void – trussed rafters carrying roof and floor load
(3) Attic – trussed rafters carrying most of roof load with separate joists for floor loads
(4) Attic – trussed rafters carrying floor and roof loads

(1) For special trusses designed to provide an unobstructed roof void for some part of the roof, the truss loading is as for standard trusses. In between the trusses, independent joists are installed, taking support on the same wall plate as the trussed rafter and on some intermediate load bearing wall (see fig. 5.15). The timber sections used in the trussed rafters will be standard and not the heavier sections required with the full attic, thus the cost of the truss is not greatly increased. Care must be taken to add the necessary herringbone strutting, and the strutting indicated in the illustration on the intermediate wall plate to brace the infilling floor joists. This particular

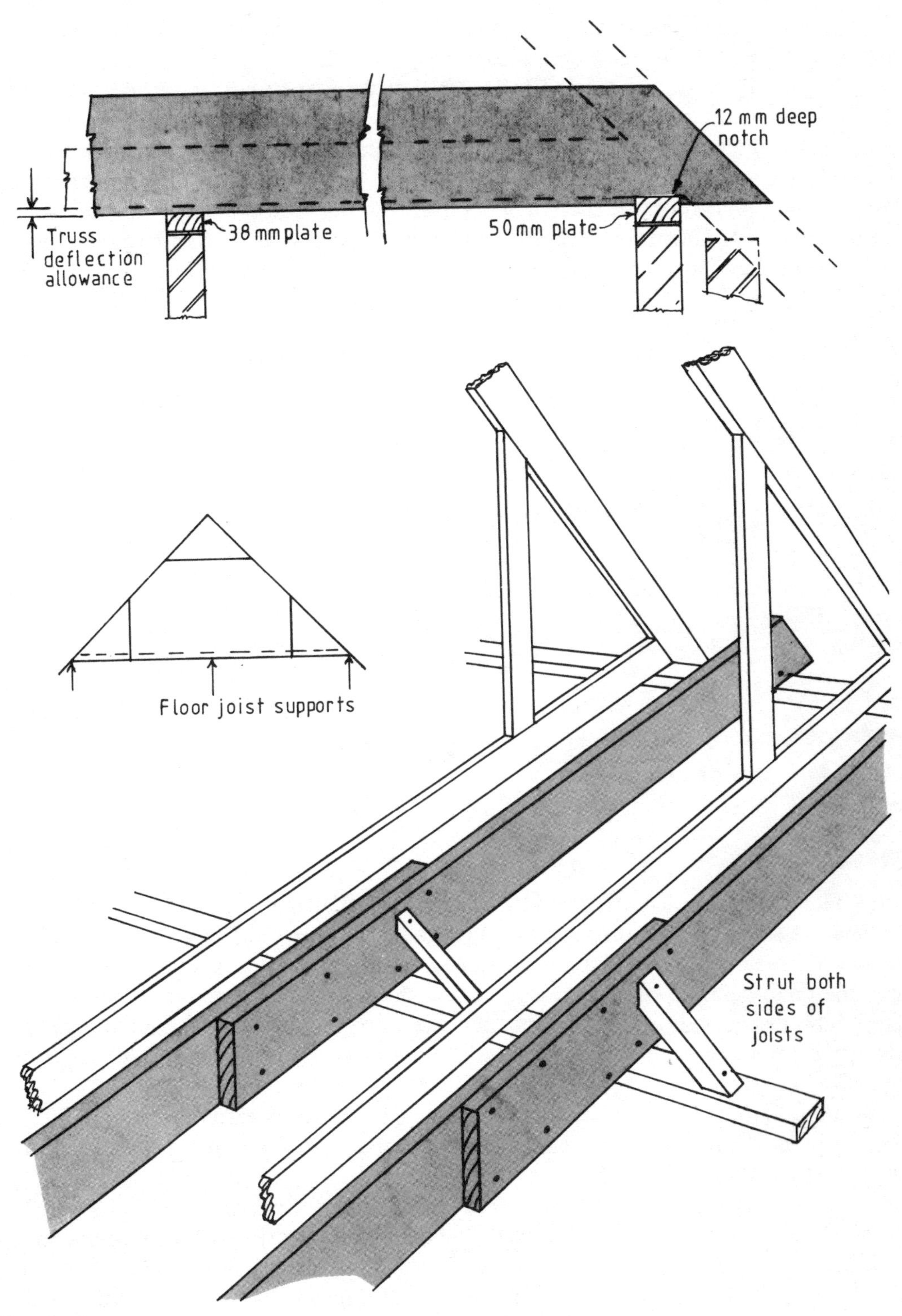
12 mm deep
notch
Truss
deflection
allowance
38 mm plate
50 mm plate
Floor joist supports
Strut both
sides of
joists

Fig. 5.15

construction is recommended where an existing trussed rafter roof is being used to carry more than simple loft storage loading.

(2) A specially designed trussed rafter, designed to carry both roof and floor loads, with or without internal support for the floor. To avoid the truss becoming almost an attic, thus containing costs, the loaded area of the loft should be kept as small as is practical, particularly on the clear span design. In both cases, the trussed rafter designer may seek to close the trussed rafter spacing down to 450 mm or 400 mm centres. Fig. 5.16 illustrates this loft option. If insulation is to be fitted between the rafters, consideration must be given to ventilation – see Chapter 7.

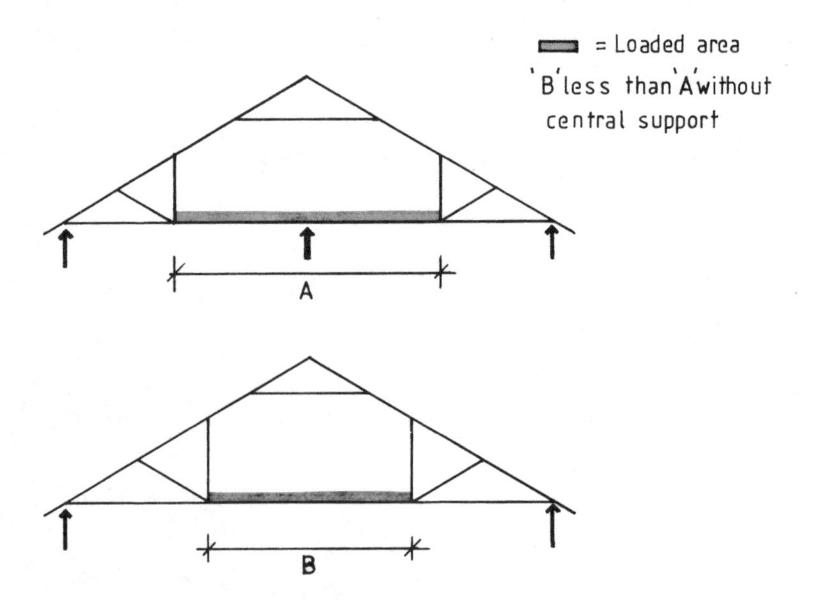

Fig. 5.16

(3) The attic design solution, given below, is particularly suited to dwellings where the span is large but the distance between gable walls is relatively narrow. Choosing the shortest span for a roof structure usually results in greatest economy. This design uses the cross-walls to support purlins onto which trussed rafters and floor joists are fixed. The design also allows the use of wide dormer windows without the need for heavier trimmers, and simplifies forming openings for stairwells. It is, however, labour intensive compared to option (4). Fig. 5.17 illustrates option (3).

(4) Fig. 5.18 shows a fully independent attic trussed rafter construction. The timbers used will now be much larger in section than for standard trussed rafters, thus increasing the unit weight; this must be considered for handling reasons. The steeper pitch necessary to allow satisfactory room within the roof can mean that the height of the truss will be outside both manufacturing and transportation practical limits, and in such instances the roof truss may be split horizontally into two components, the split occurring at the ceiling joist or collar position. Producing two trussed rafters to form each frame therefore adds to the cost of such construction. The overall economy of the

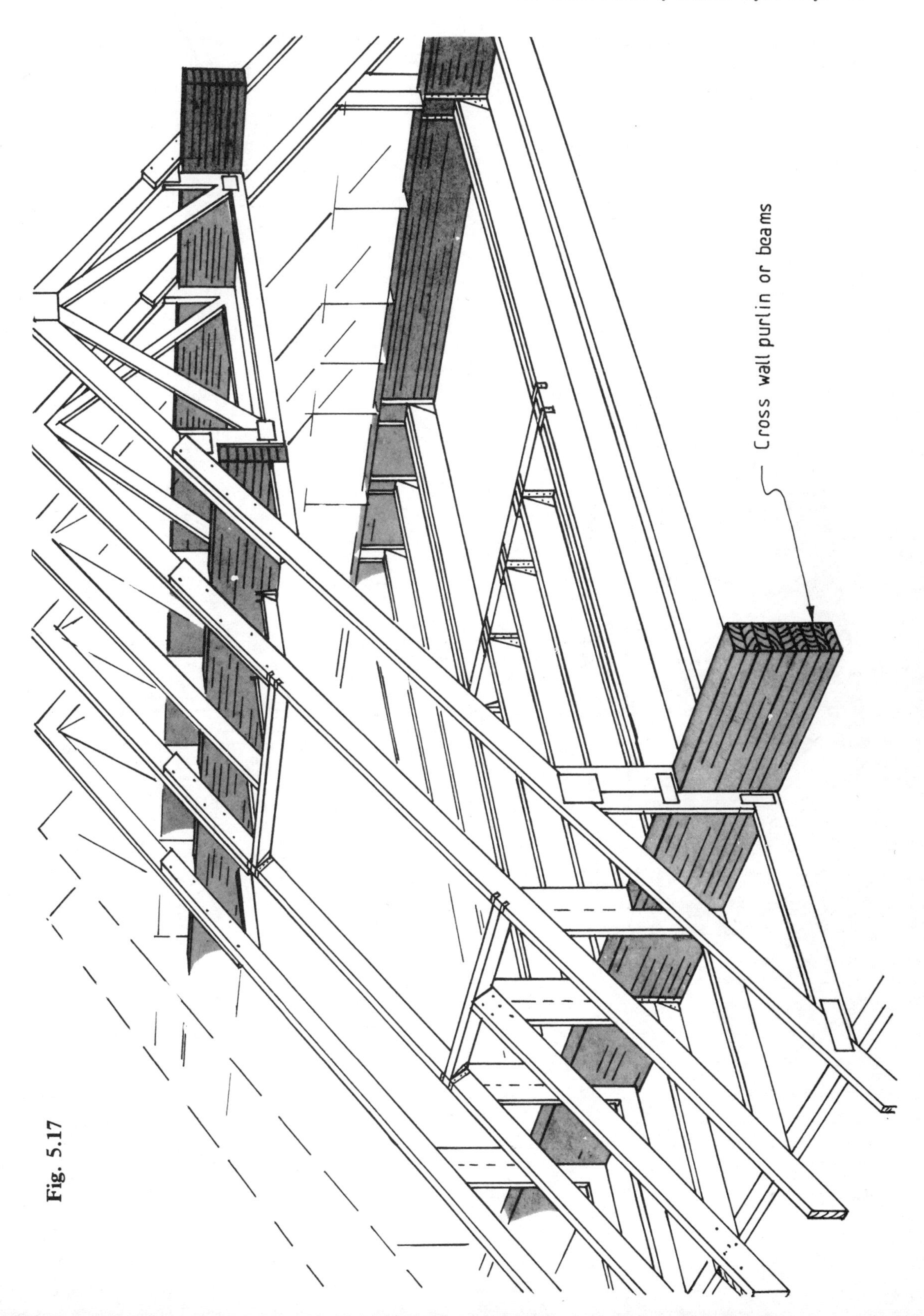

Fig. 5.17

Fig. 5.18

attic trussed rafter roof is very dependent on dormer or roof window and stairwell planning, this aspect is now discussed.

If a clear spanning attic trussed rafter is to be used then there is complete freedom of roof planning, both at attic level and on the floor below. However, location of stairwell and dormer or roof windows can have a dramatic effect on the cost of the attic structure. Each trimmed opening will require girder trussed rafters, the wider the opening the more trussed rafters to each girder. Bearing in mind that each attic trussed rafter will cost approximately four times its standard non-attic equivalent, it can be seen that girders must be kept to a minimum. Set out below are five basic rules of attic trussed rafter roof design economy.

(1) Plan to keep trussed rafters on a 600 mm spacing
(2) Plan the stairwell to run parallel with the bottom chord and keep as narrow as possible.
(3) Plan dormer and roof windows as narrow as practical (some roof windows will fit between 600 mm centre trusses)
(4) Plan dormers and windows in-line across the building
(5) Attempt to keep the overall height of the truss within transportable and/or manufacturing limits. This is often around 4 m, avoiding the additional cost of having to split the truss into two sections. Fig. 5.19 illustrates a split truss.

Fig. 5.19

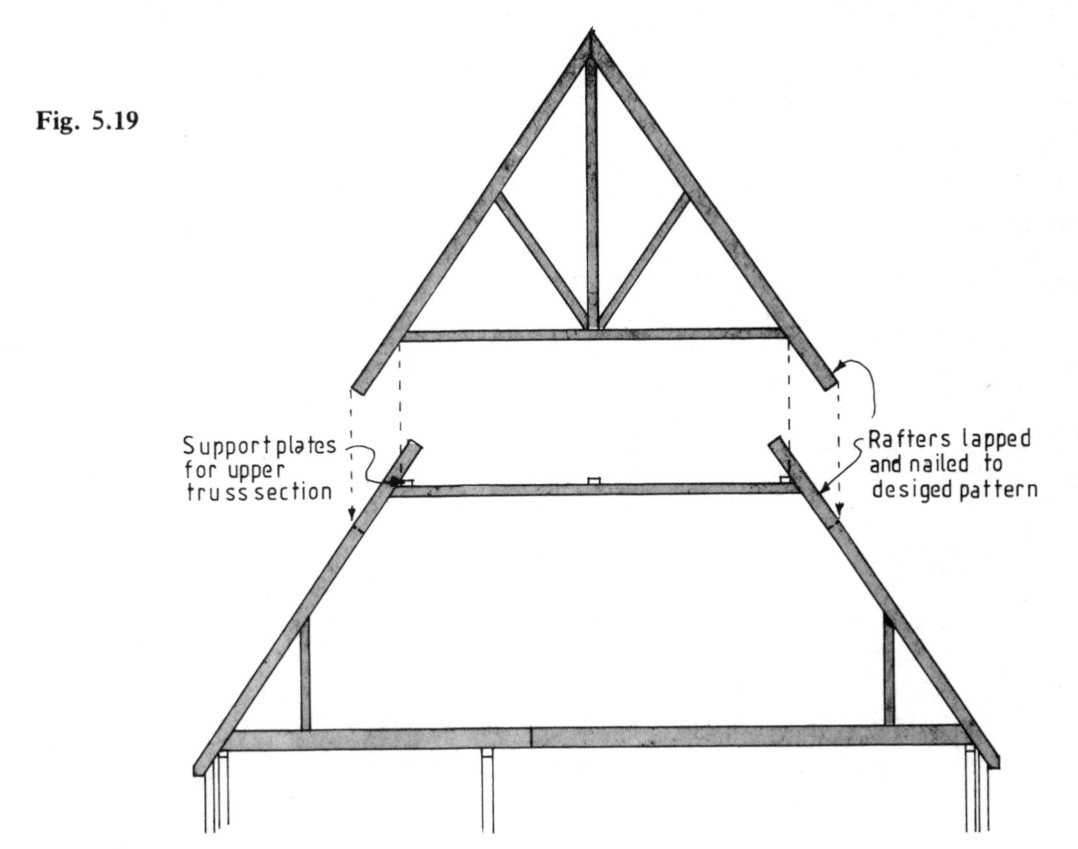

Fig. 5.20a illustrates a poor attic layout involving many heavy girders, few trussed rafters at 600 mm centres, and much site fixed infill and should be avoided. Fig. 5.20b, however, is a much more economical layout involving lighter girders, more trusses at 600 mm centres and less site infill. The layout in fig. 5.20a could be solved

Fig. 5.20a

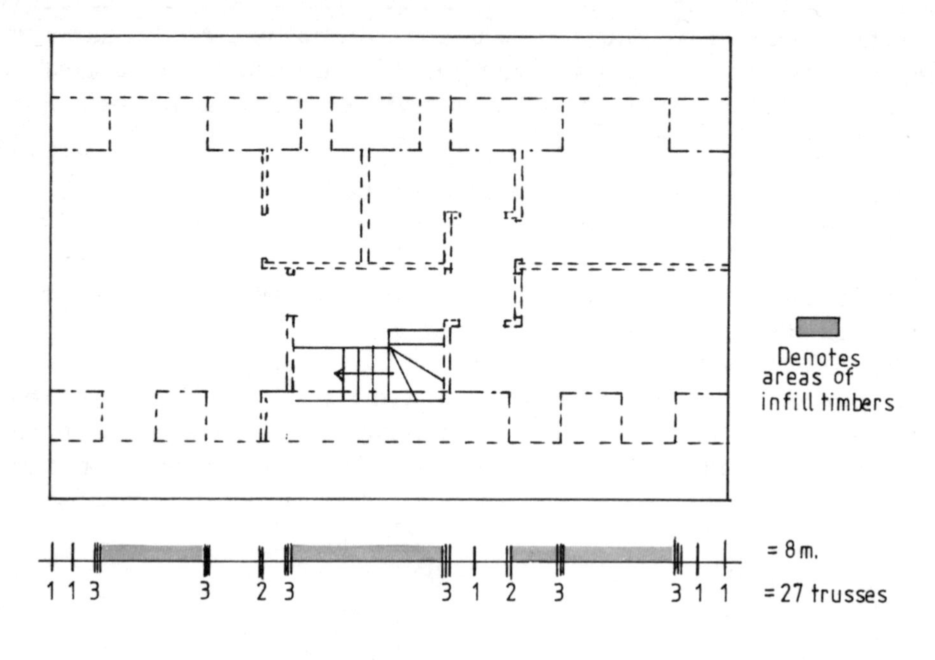

Fig. 5.20b

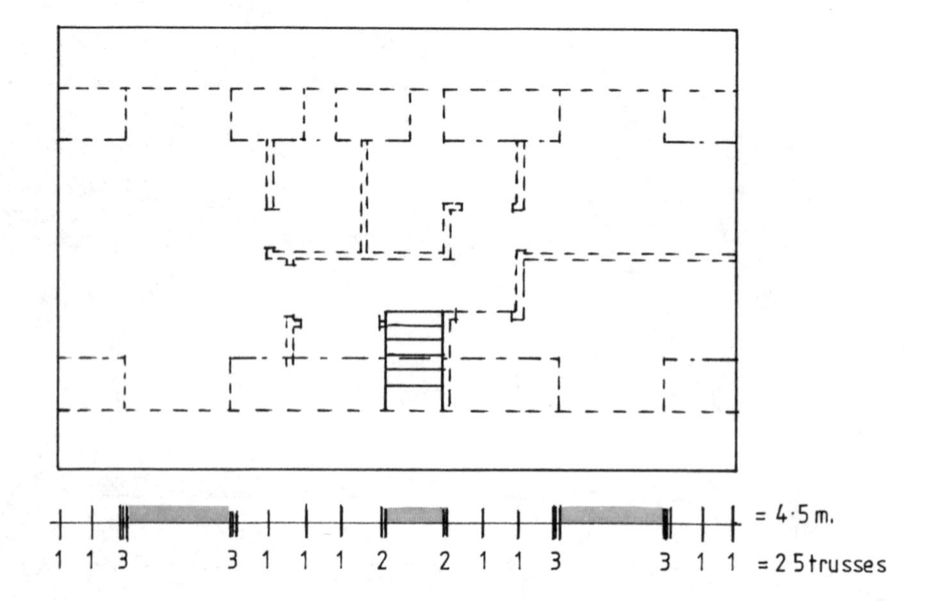

using the trussed rafter, purlin and joist design shown in fig. 5.17. Intermediate support may be required for the purlins, with the floor joists being supported by further purlins or directly off walls below.

Further variations in attic truss solution can be found in chapter six.

OPENINGS FOR DORMERS, ROOF WINDOWS AND STAIRWELLS IN ATTIC TRUSSED RAFTER ROOFS

Openings in attic trussed rafter roofs can be formed in a similar manner to those shown in the traditionally constructed roof section, chapter three. Using two or more trussed rafters as trimmers, both infilled rafters and floor joists need to be supported from those trimming trusses. Short purlins are used to pick up the ends of the trimmed rafters, the purlins themselves being supported on girder trusses. The purlins should be supported on purlin posts built in to the attic girders at the lower purlin level, and supported on the ceiling ties as close to the joint between top chord and ceiling tie as possible for the upper purlin. The purlin should be notched under the top chords to give an adequate birdsmouth for the oncoming infill rafters. The purlins will of course carry the infill rafters, leaving the floor joists to be infilled. This can be done using short lengths of floor joists of a matching depth to the bottom chord of the trussed rafter fixed across the opening, the infill joists being supported in joist hangers fixed to the sides of the bottom chord.

Stairwell openings can be similarly constructed, but in this case the whole of both rafter slopes will need supporting on purlins as described above. Figs. 5.17 and 5.18 illustrate typical trimmed openings in two types of trussed attic construction. If the sloping ceiling in the infill area between the girder trusses need not align with the main roof, then the purlin can be lifted, and lighter smaller sectioned infill rafters used. On narrow openings between girder trusses it is practical to use joists hangers as purlin supports, fixing these directly to the sides of the top chords of the girder trusses.

As has been seen in figs. 5.20a and 5.20b, the positioning and size of openings can significantly affect the cost of constructing trussed rafter attic roofs. Attention is therefore drawn to the design in fig. 5.17, which, because it is based on large purlins, allows great freedom of creating openings of any shape or size between the upper and lower purlin without structural complications. If a large number of openings, or alternatively extremely large individual openings are required, then this option is strongly recommended.

BRACING THE ATTIC TRUSSED RAFTER ROOF

The trussed rafter attic roof must be constructed strictly in accordance with the designers' and manufacturers' drawings. It must be fully appreciated that the structure about to be constructed forms almost 50% of the new home and will carry far more load than the conventional roof structure. The trussed rafters impart great rigidity across the building within themselves, but rely purely on the bracing construction between them for their lateral stability, and therefore the lateral stability of the whole building above the wall plate position. It is *not* correct to consider that brick and block gable ends will impart any support for the roof structure in this lateral direction, indeed the timber roof structure itself must be strong enough to restrain adequately the large gable end areas.

A preliminary consideration of bracing is given below, but the more detailed implications of correct bracing are dealt with in chapter seven where they cover both traditional and trussed rafter attic roofs.

Temporary bracing of these larger heavier trussed rafter components is vital. The temporary diagonal brace placed on the top of the top chord should be at least 22 mm × 97 mm in section and well nailed to both plate and the trussed rafters as far up the rafter as practical. On a single-storey building it may be possible to place a temporary support, or prop, from the floor slab to the vertical side wall and top chord junction of the attic trussed rafter for additional security. Erection should proceed basically as for the standard trussed rafter roof, except that temporary diagonal bracing should be added to the underside of the top chord until permanent bracing of the roof has been achieved. This temporary bracing may well have to be fixed on the underside of the top chord or rafter within what will become the room area.

Longitudinal binders should be fitted generally as before, but there will be more of them and these are indicated in fig. 5.21. This illustration does *not* show all of the bracing required in an attic roof, further reference should be made to chapter seven, fig. 7.22. The floor boarding will of course eventually act as a substantial binder for the bottom chord, therefore only temporary binders need to be installed prior to the laying of the floor itself. These binders are essential if correct spacing of the trusses is to be maintained, which will of course assist cutting and fitting of the floor boards at a later date.

Fig. 5.21

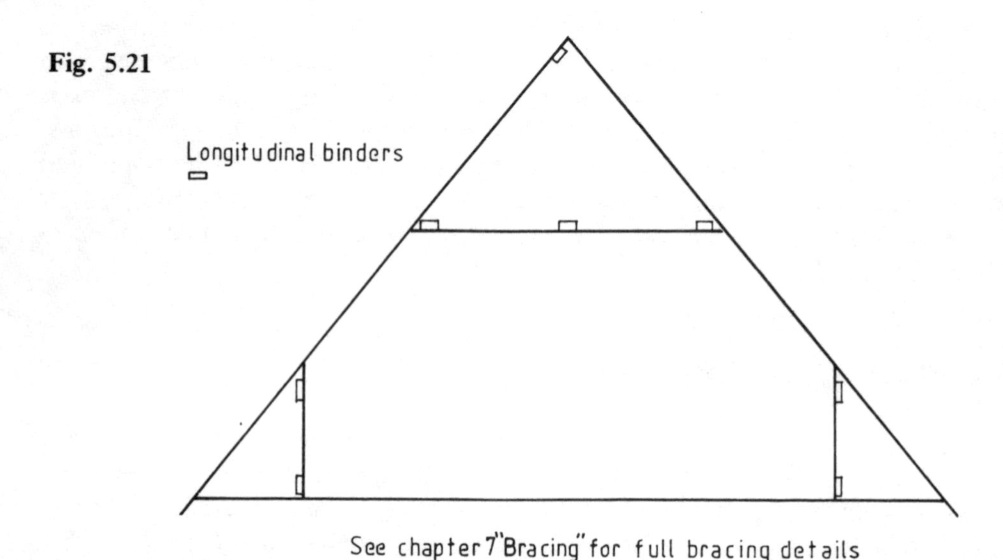

Fig. 5.22

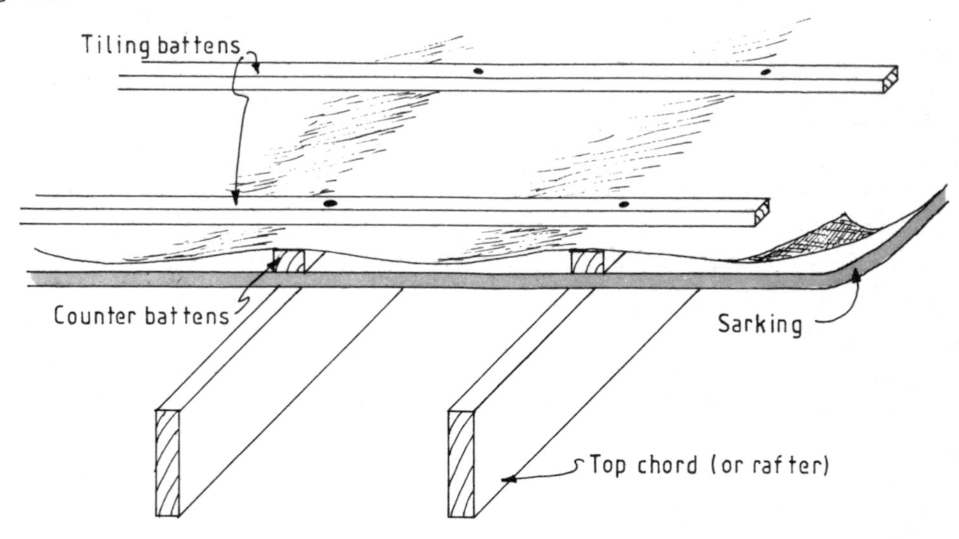

The problem of the lateral stability of the roof remains to be solved on a permanent basis. In Scotland, Scandinavia and the USA where roof sarking is common, this forms a substantial brace and nothing further is required. The sarking often takes the form of a sheet material nailed to the top of the rafters over the whole roof area (see fig. 5.22). BS 5268 part 3 and No 1 in appendix A Note 4, gives guidance on suitable sarking materials and fixing.

TRUSSED RAFTER SHAPES

Other common trussed rafter shapes are shown in fig. 5.23.

A Fink – equal pitch
B Stubb fink or bob-tail
C Cantilever fink
D Northlight
E Horizontal split truss
F Monopitch
G Asymmetric – unequal pitch
H Inverted fink
I Raised tie
J Horizontal split attic

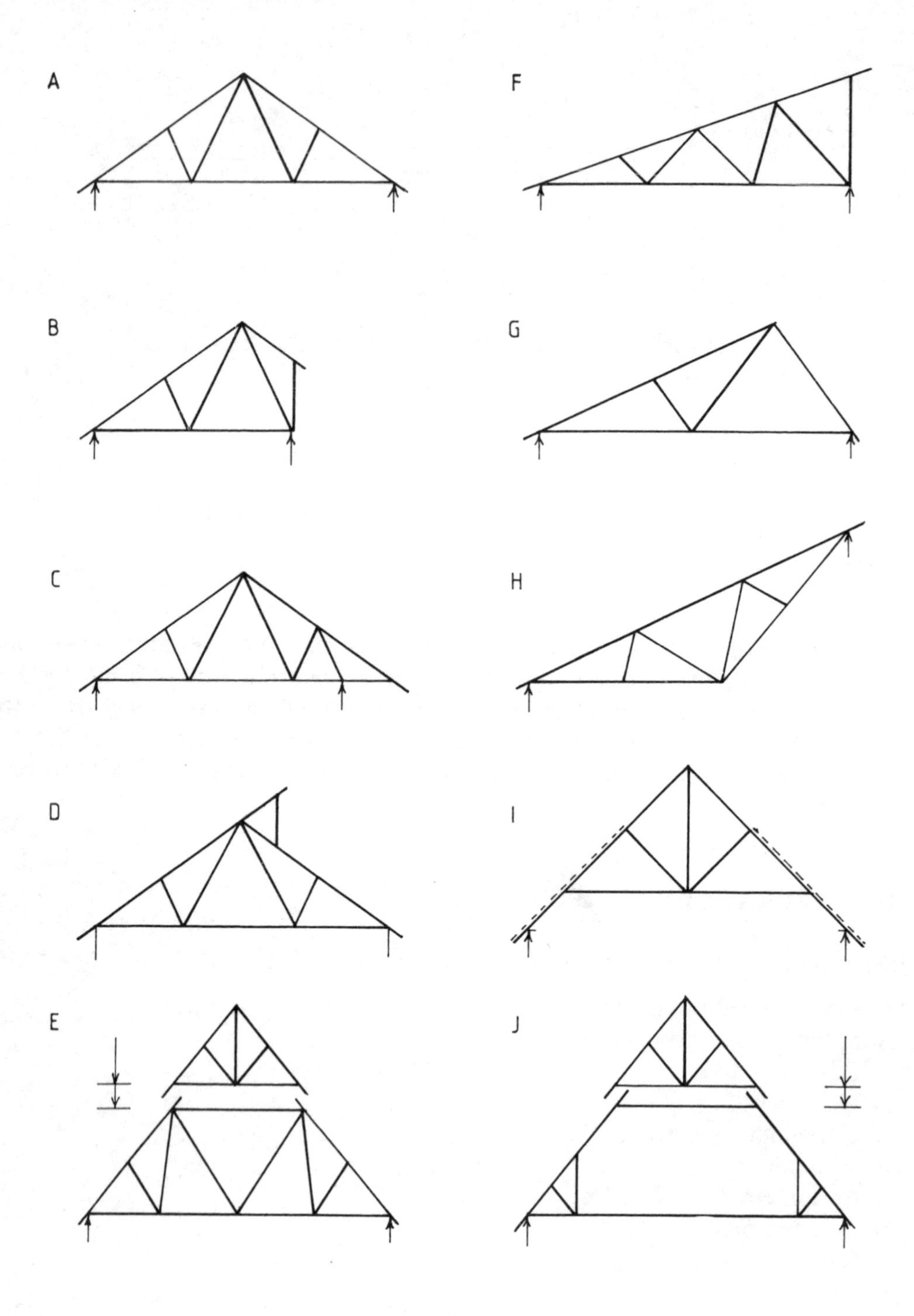

Fig. 5.23

CHAPTER 6

Truss plate systems – variations in design

TRUSS PLATE SYSTEMS

Most users of trussed rafters will first come into contact with the name of one of the plate system manufacturers on receipt of a set of computer printed calculations. If these are not asked for, it is possible that the user may never know which brand of plates has been used in the manufacture of the trussed rafters supplied to him.

The plate manufacturers or 'system owners' supply both plates and design information to trussed rafter manufacturers who clearly are free to choose which system they prefer. The choice will be made on plate price, but more importantly and increasingly a deciding factor, on the degree of design information and technical back-up available from the system owner.

Whilst at first sight the type of plate used on the trussed rafter may seem of little importance to the end user, because his contract to purchase trussed rafters is with the trussed rafter manufacturer, it can affect both the quality of the service and product supplied, and more importantly the precise construction form used on the roof.

SYSTEMS AVAILABLE

This brief examination of the systems currently available in the UK is not intended to be a basis on which to make a choice of system to be used for production, but is intended to give an indication of the differences in approach to design and provision of design information by the five plate system manufacturers. These manufacturers in alphabetical order are Bevplate, Gang-Nail, Hydro-Air, Trusswal Systems and Twinaplate. Gang-Nail and Hydro-Air vie for the title of 'brand leader' and certainly between them hold the lion's share of truss plates produced and sold in the UK, and probably world-wide. All truss plate manufacturers belong to the International Truss Plate Association. British Standard 5268 part 3 sets out minimum standards for plate specification and all manufacturers hold Agrément Board certificates for their plates.

All produce plates made from galvanised metal strip and, if greater rust protection is required, all can produce stainless steel plates.

Needless to say all plate manufacturers provide the design information necessary for individual truss shapes, mostly through a computerised system. All now have solutions for hip construction to their individual standard layouts on computer program. The computer programs provided for use, usually on microcomputers at the manufacturing station, vary greatly in their sophistication – from Gang-Nail's, which will not only design the truss but also provide a computer drawn layout of the roof members, to those of Bevplate, Trusswal and Twinaplate which design the shape only.

Hydro-Air encourage the manufacturer to use their centralised design service for more complex structures, when again computers are used both to print the engineering information and to draft the drawing layouts. All provide structural engineering back-up for their manufacturers. Whilst Bevplate, Trusswal and Twinaplate do not appear to encourage their manufacturers to offer 'whole roof' packages, that is including binders, bracing, ties, hip and valley infill, etc., Hydro-Air seem to have taken the opposite approach, and to back this up have produced an extremely wide range of the necessary steel work to connect the trusses together, build them into brickwork and fix the wall plates. Gang-Nail do not actively encourage their producers to carry out whole roof designs unless they feel that their manufacturers' engineering expertise is adequate to do so but themselves offer a full design consultancy service.

HIP ROOFS

Bevplate

On hips up to 6 m span a double fink truss is placed at the hip peak position, with normal fink trussed rafters being placed at the traditional 600 mm centres for the roof containing the ridge. In the hip area, this is infilled using traditional construction techniques as set out on the detail sheet provided to the trussed rafter manufacturer. For larger spans, two alternative designs are available. Firstly a girder truss of howe configuration is used located at hip peak position, which supports a set of monopitched trusses at right angles to the girder. The monos are arranged in doubles to form girders at quarter span points of girder truss (to coincide with the howe quarter point members) with the remainder infilling between them, and with a full mono pitch configuration truss located on the main roof ridge line. The mono girders support site infill or further small monopitch trusses with extended top chords cut at the hip rafter intersection. Fig. 6.1 illustrates this design.

The second Bevplate solution uses flat top trusses in the upper half of the hip area, the truss at the hip peak position being a single truss from the main roof area where a ridge exists. A flat topped girder truss is used part way down the hip slope, to support monos wih extended top chords, the chord extensions being cut to fit the hip rafter itself. The extreme lower corners of the hip are site infilled using traditional methods. Again a standard detail is available for this construction.

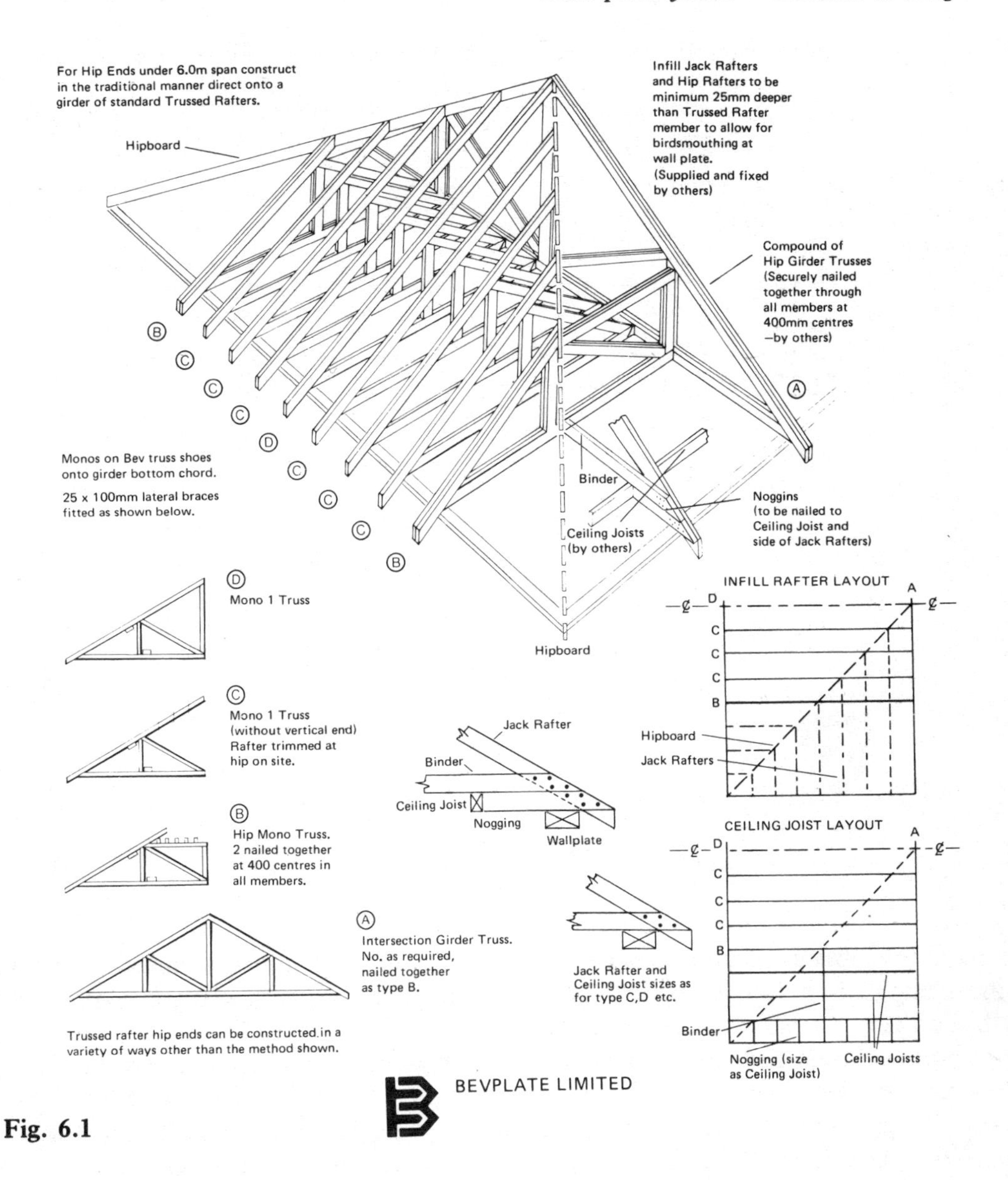

Fig. 6.1

Gang-Nail

Gang-Nail have a range of hip solutions available to the trussed rafter manufacturer via their computer programs. On hips with a span of less than 4 to 5 m, Gang-Nail allow the use of site constructed infill using traditional methods. This site construction is supported on a girder truss placed at the hip peak position.

The most commonly used Gang-Nail hip solution is that based on a fink truss and is illustrated in fig. 6.2. The term 'fink based' is used because the girder trusses have an internal member configuration corresponding generally to the members of the fink

Fig. 6.2

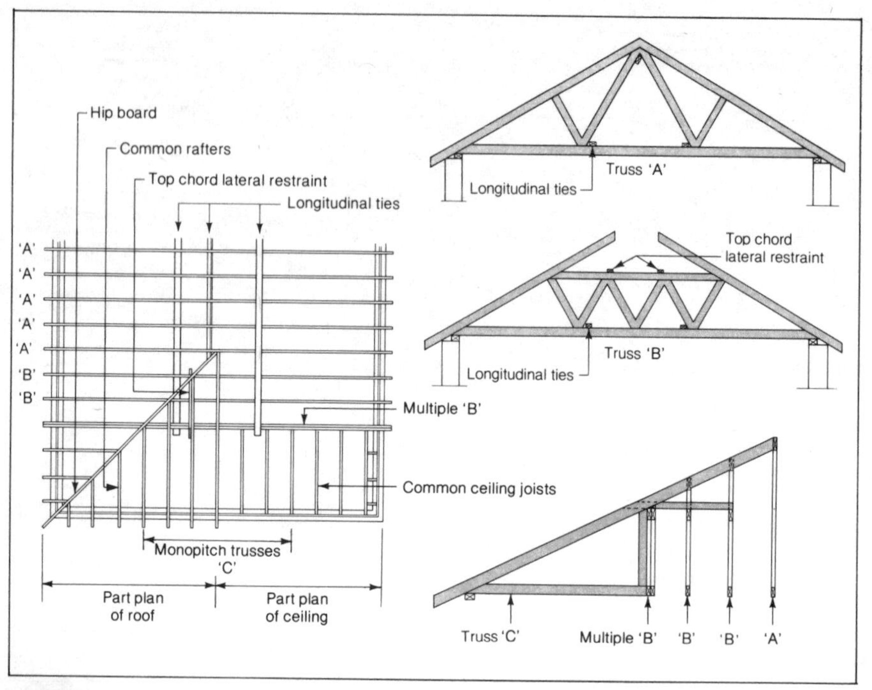

A Redland Company

Gang-Nail Limited

truss. This facilitates continuity of the location of binders from the main roof, through into the hip end at the node points on the bottom chord of the fink, aligning precisely with the node points on the bottom chord of the hip girder trusses.

The method uses a single standard fink truss at the hip peak position, with a girder truss comprised of two or three plies maximum, at a point approximately half way down the hip slope. Between the girders and the individual fink are placed further single ply flat topped hipped trusses which have extended top chords. These top chords are cut on site to fit the hip rafter itself, thus automatically infilling the upper part of the hip without further site infill. The flat topped hip girder itself supports monopitched trusses which, unlike the other systems, are supported on their top chords on the top of the flat top hip girder. The top chords of the monos are again extended at the factory and cut to the hip rafter on site. This particular method leaves only a small area for site infilling at the two lower corners of the hip itself.

The other hip solutions available from Gang-Nail are options available through their concept two version three computer programs. The fink based hip, discussed above, results in non-standard spacing of the intermediate hip trusses, and to overcome this feature which is occasionally objected to by builders, a 'standard centres hip' is available. This places the intermediate hip trusses at centres equal to those of the main roof truss, with the girder truss being positioned accordingly. Monopitch trusses complete the hip in a similar manner to the fink based hip.

It is sometimes beneficial from a manufacturing point of view, to control the positioning of the hip girder by fixing the span of the monopitch extended top chord trusses. Again the construction of the 'standard set back hip' is similar to that of the fink based hip but the monopitch span becomes the design priority, locating the girder truss and the intermediate hip trusses as second priority. In the last two systems described, the monopitch trusses are supported on their bottom chords on truss hangers, unlike the fink based hip where the monos are top chord supported.

The final hip solution is that known as the 'girder based hip' which uses a howe or similar girder. This particular layout is very similar to that given in fig. 6.1, but a further economy in production is achieved by Gang-Nail's design in that trusses B, C and D from fig. 6.1 combine into one truss similar to truss type B, but with an extended top chord. Monopitch trusses are bottom chord supported in truss shoes on the double trusses 'B', in place of the site fixed jack rafters of fig. 6.1. This design is also similar to fig. 6.6.

Hydro-Air

Hydro-Air offer no less than four solutions to the hip construction problem. On roofs of up to 5 m span a double fink truss is used at the hip peak position with a traditionally cut and fitted roof to infill the whole of the hip area. Interestingly their design sheet for this construction gives detail of the hip rafter to double fink girder connection. It also highlights the problem of eaves and fascia detail where the hip pitch differs from that of the main roof, and stresses the need to tie the wall plates together at the hip corner.

Fig. 6.3 shows the Hydro-Air series one prefabricated truss hip solution using a combination of decreasing height flat topped trusses, to a point approximately half way down the hip where it locates a flat topped girder truss. This girder in turn supports monopitched trusses in truss shoes on the bottom chord. The monos are then extended up to meet the hip rafters by site fixed infill, and the extreme lower corners of the hip are again infilled traditionally.

The series two and three hip solutions use a set of flat topped trusses but, unlike series one, these are all of the same height. The principal advantage over series one is that as all flat topped trusses follow the same profile it is a cheaper roof for the manufacturer to produce, but conversely it has a greater proportion of site infill to complete the roof. On the series two solution the offcuts from the extended top chord of the monopitched trusses are used to infill the rafter area above the hip truss. Fig. 6.4 illustrates in plan form the series two hip solution.

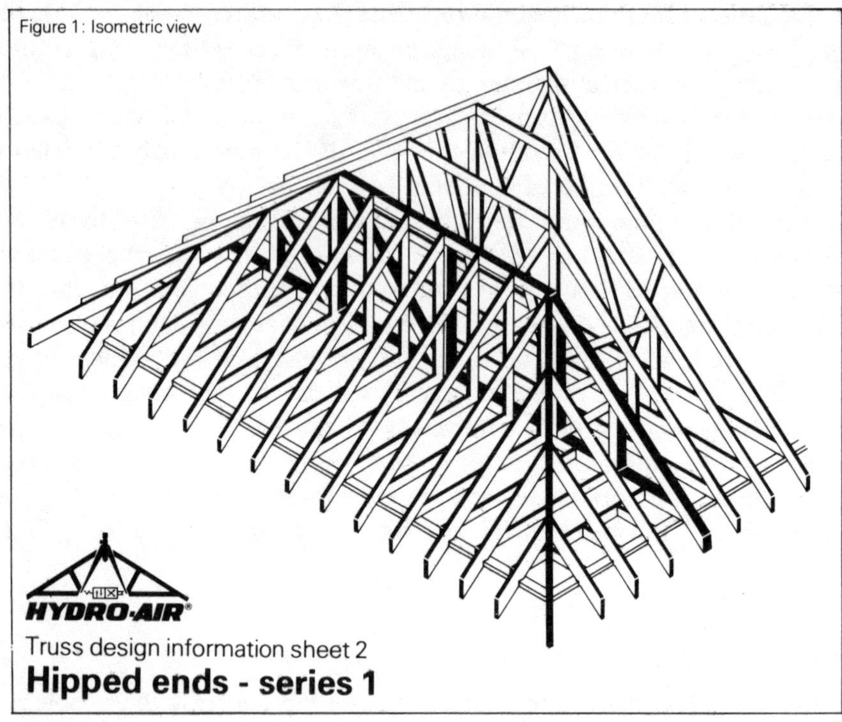

Fig. 6.3 (part of sheet 2 shown above)

Hydro-air hipped end series 2 (part of information sheet No. 3)

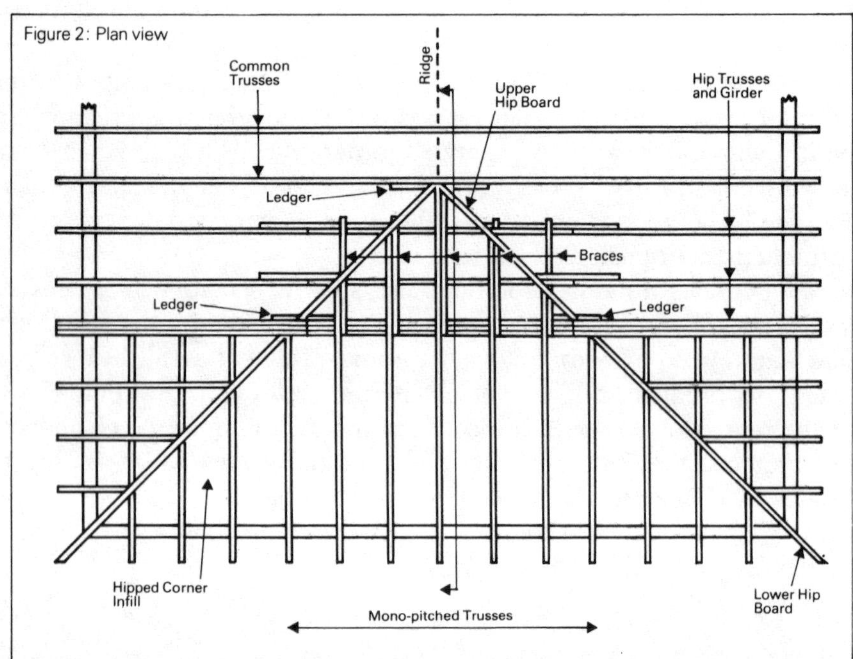

Fig. 6.4

Trusswal

Trusswal have three hip solutions, two of which are based on their 'step down' system. The first offers the maximum prefabricated components of all manufacturers, with only the hip rafters themselves being loose timbers. As can be seen from fig. 6.5, a series of trusses is produced from one being a full height stepping down the hip to one about half way down. Monos are then supported on the lowest hip truss, the extreme lower corners being filled with simple prefabricated jack rafter and ceiling tie units. The second solution, given in fig. 6.6, involves placing a girder truss at the hip peak, supporting special trusses with extended top chords to give the hip shape. Further side prefabricated jack rafter and ceiling tie assemblies are then fitted, again making a completely prefabricated hip solution with no site infill.

Fig. 6.5

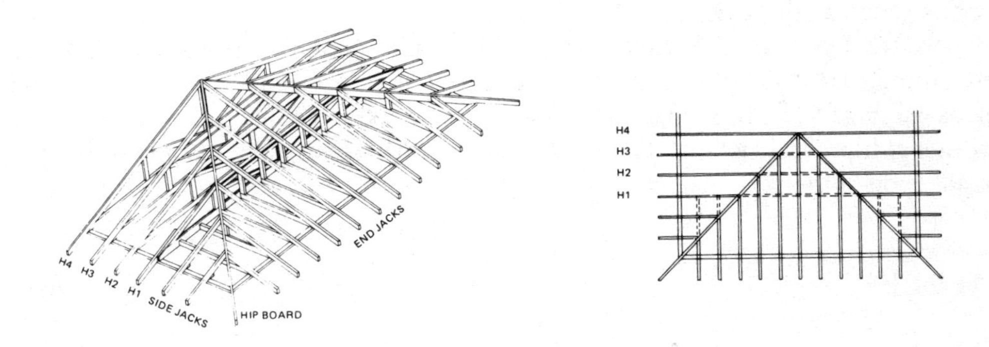

TRUSWAL STEP-DOWN HIP SYSTEM

Fig. 6.6

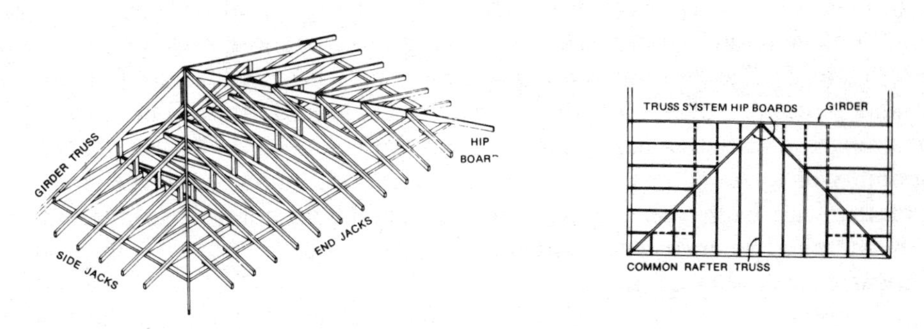

TRUSWAL TERMINAL HIP SYSTEM

The third solution is similar to the first described in the Gang-Nail section, but the positioning of the hip girder and its configuration are not determined by fink geometry. The top chords of the flat top trusses and the monos are not extended to meet the hip rafter and therefore the method requires slightly more on-site infill to complete the roof structure.

Twinaplate

Two systems exist, the first providing a highly prefabricated hip, the second providing the main structural members only for what is essentially a traditionally constructed hip. The first method, illustrated in fig. 6.7, uses a hip truss girder with intermediate trusses with extended top chords between the girder and the hip peak. Monos with extended top chords fill the area between hip girder and the hip end, plus two small king post trusses fitted to partly infill the extreme lower hip corners. Site infill timber is used to complete the hip.

The Twinaplate technical guide emphasises the need to fit binders within the hip area and to fit struts under extended top chords where their length exceeds certain limits set out in the guide.

The second system generally consists of a double standard truss at hip peak position and a hip girder truss at mid-point between peak and hip lower end. These two girders act as supports for hip boards, the jack rafters and ceiling joists of essentially a traditional hip construction. The guide states clearly this system is only to be used for 'small' roofs although small is not quantified in terms of either pitch or span.

VALLEYS

The construction to be examined is that required to provide a roof on a T-shaped building where there is no supporting wall for the main roof at the T junction. Fig. 6.8 illustrates the roof in question. Valleys follow a similarity of designs between the various manufacturers, but important constructional variation still exists.

Bevplate, Gang-Nail, Hydro-Air and Trusswal: all of these systems use a girder truss at the T junction to carry the ends of the standard main roof trusses. This girder is usually of a howe or similar configuration giving three or more node points to support the bottom chord. The bottom chord is of course very heavily loaded by the oncoming trusses and is usually of greater depth than the standard truss. This provides both greater strength and better fixing for the steel truss shoes. Hydro-Air, concerned by the possibility of torsional or twisting effects on the bottom chord of the girder caused by all of the load from the common roof trusses being applied on one side of the girder, recommend a construction illustrated in fig. 6.9 to control this effect on girders where spans exceed about 8 m.

Having constructed the intersection of the two main roof slopes, the valley is infilled using prefabricated reducing trussed frames. Gang-Nail use a valley board laid in one continuous length on top of the main trusses, with one of the standard main roof

HIPPED END PLAN

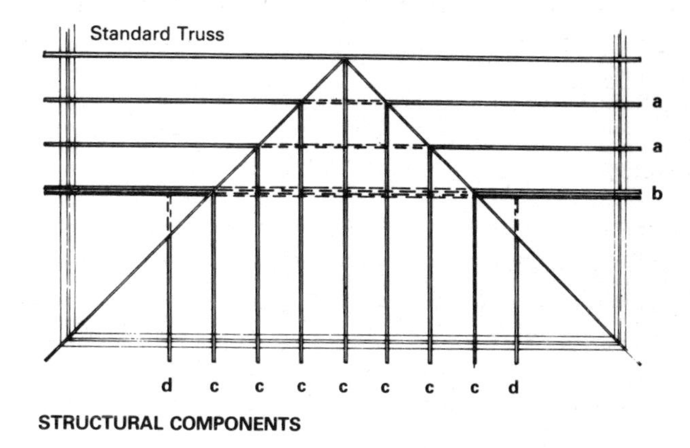

STRUCTURAL COMPONENTS

TWINAPLATE

HIPPED END CONSTRUCTION

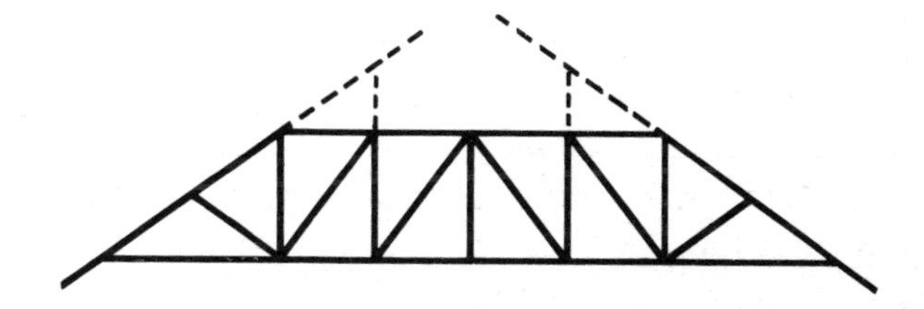

(a) SINGLE GIRDER UNITS WITH EXTENDED RAFTERS

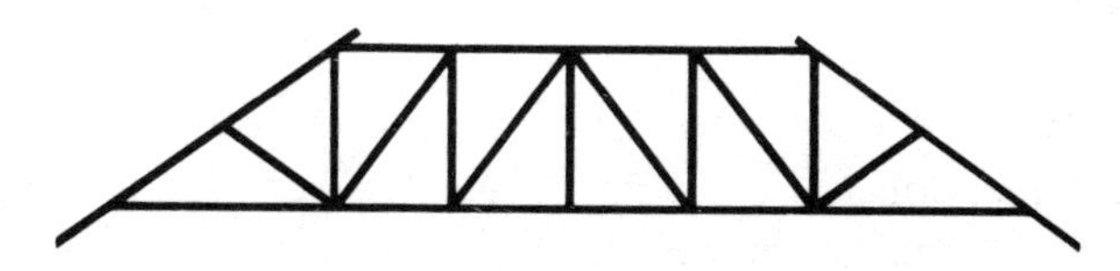

(b) MULTIPLE GIRDER UNIT

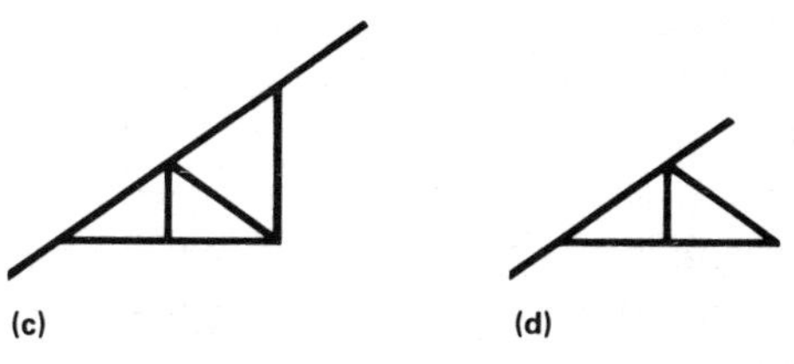

(c) (d)

MONO PITCH AND KING POST WITH EXTENDED RAFTERS

Fig. 6.7

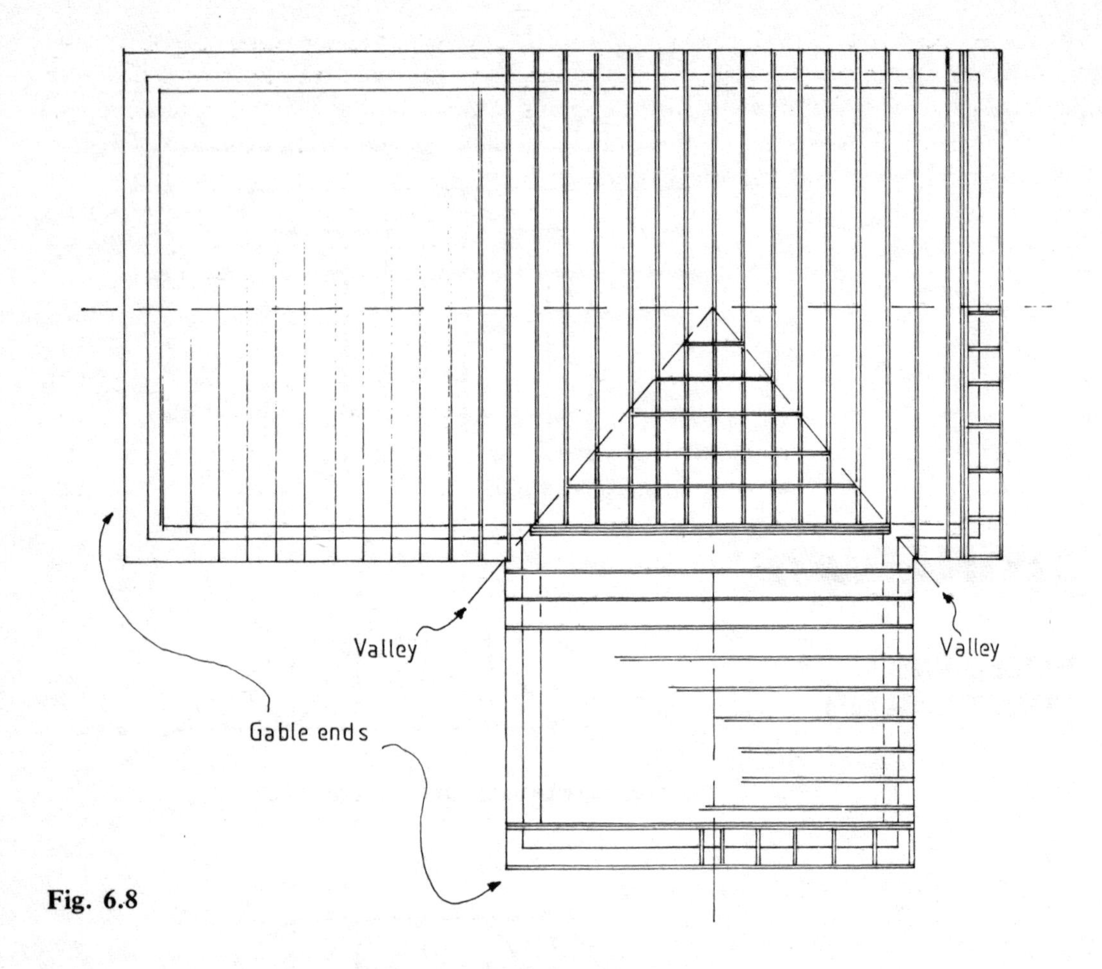

Fig. 6.8

Part of Hydro-Air design information sheet 5

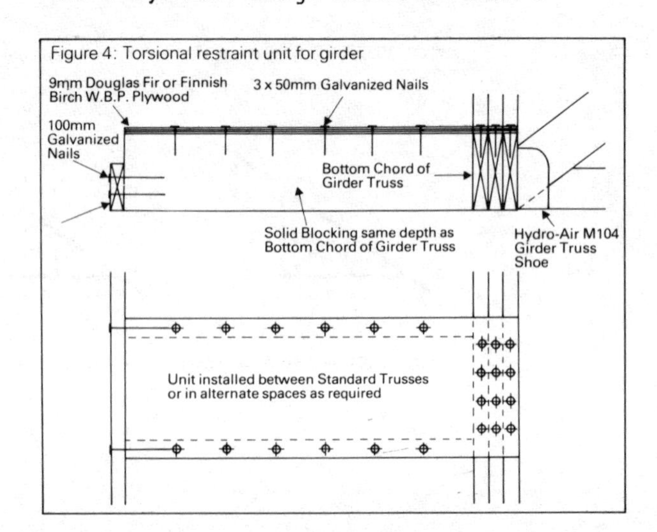

Fig. 6.9

trusses having been placed at the centre of the valley ridge position. Most systems show the uprights of the valley infill frames at 1.2 m centres. Bevplate, however, illustrate these at 600 mm centres, thus giving support to the infill truss top chord at each main truss it passes and more uniformly distributing the load on a roof below. Hydro-Air stipulate that the tile battens shall be fixed on the main roof top chords between the valley frames, thus providing the bracing assumed in the main roof truss design. Hydro-Air also show the valley boards as short noggins fitted between the top chords.

The valley reducing truss set must either have the underside of the bottom chords cut to fit the slope of the truss on which they sit, or be fixed on a separate triangular batten nailed on top of the common rafter top chords to give a square seat for the valley frames. The reducing trusses are located precisely so that the heel joint rests on the valley board, which is in turn supported by the top chord of the main roof truss. The bottom chord of the valley set must be nailed to each top chord of the truss supporting it, thus replacing the tile batten which would normally be in that position. Apart from Hydro-Air, as mentioned above, the other systems do not illustrate a tile batten between the valley frames which are normally placed at 600 mm centres. Fig. 6.10 illustrates the valley detail shown in the Gang-Nail design manual.

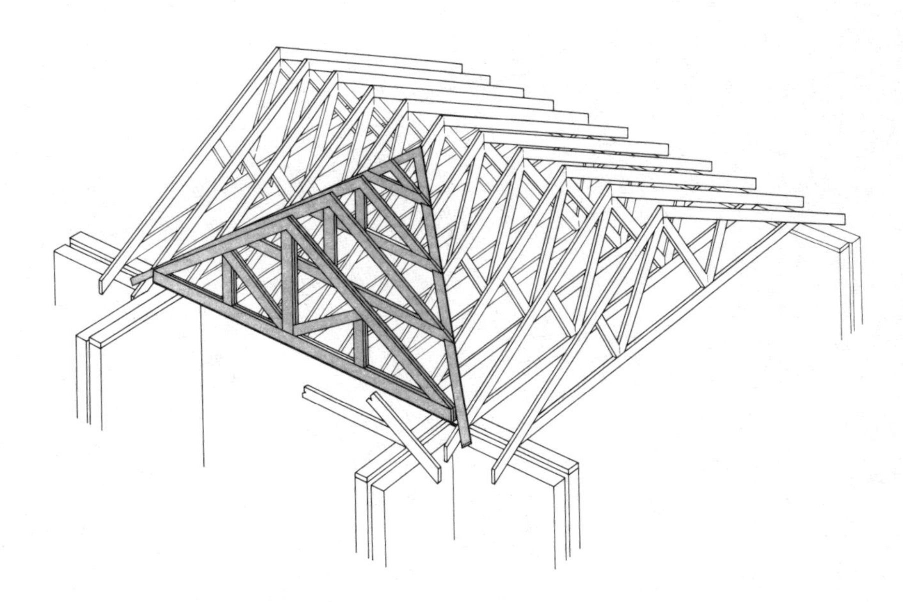

Fig. 6.10

Gang-Nail positively discourage the use of traditional valley infill, yet Trusswal are satisfied with a traditional valley board, jack rafter and ridge solution. Trusswal make it clear that if traditional means are used the additional bracing must be fixed on the top chords under the valley area to replace the tile battens as mentioned above.

Twinaplate

Similar to the above systems, Twinaplate's 'T' valley solution again supports the main roof from a girder truss. In this instance, however, Twinaplate illustrate a fink girder rather than the howe used by most of the other systems. Fig. 6.11 illustrates the Twinaplate valley, which again shows a valley set, but this differs in that the location of the valley frames is not necessarily at the point where the valley strikes the top chord of the main roof. The internal vertical members of the valley set are, however, at 600 mm centres and are located to coincide directly above their top chord main roof supports. The underside of the bottom chord of the valley set is cut to suit the angle of the main roof and must be nailed to each top chord it passes. Attention is drawn to the need for temporary bracing of the valley set until the tile battens are fixed, thus ensuring the frames remain truly plumb.

Fig. 6.11

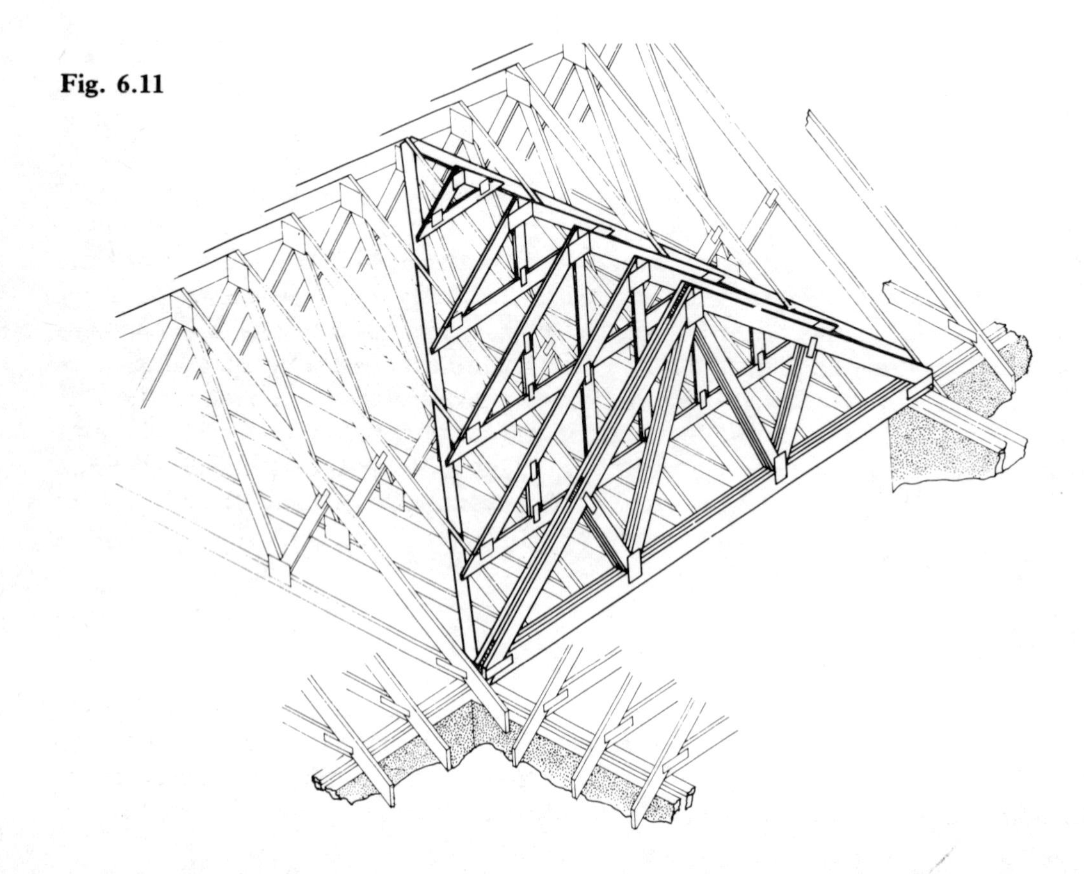

Fig. 6.12

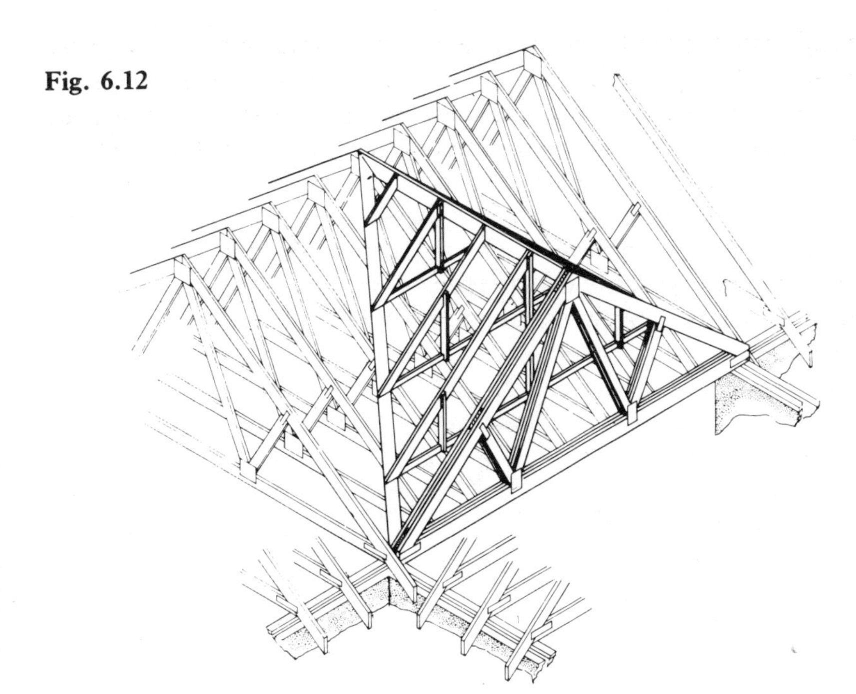

Fig. 6.12 shows Twinaplate's site constructed solution, their technical guide giving typical timber sizes to be used. In this solution the valley lay board is placed on top of the main roof trusses in the normal manner in one continuous length, with bevelled bearers again to match the main roof pitch being fixed across the main roof trusses (at presumably 600 mm centres on plan), both to tie the trussed rafters together and to form a support for the valley common rafters. A note is made that with both of these constructions, counter battens may be fixed to the underside of the ceiling tie in the room where the compound truss occurs. It is suggested that this will eliminate cracking which may appear in the ceiling due to shrinkage and differential deflection. Anticipating a different height to the rooms in this area of the building, the note continues that allowance must be made for the counter batten thickness when the trusses are initially fixed, presumably by adding an additional plate of thickness equivalent to the counter battens on top of the conventional wall plate. It would appear that this could have the effect of making fascia alignment difficult.

ATTIC TRUSSES

All systems produce attic roof truss design solutions, but on these the shape and layout can not, of course, be varied greatly. The overall height of the attic truss may well dictate that a two part horizontally split truss shall be used as discussed in chapter six, this being made necessary possibly because of the production system being used having limited height jig dimensions, or more likely by transportation limitations. A 4 m high truss is often the maximum transportable because of the height of telephone wires. Hydro-Air, however, seem to have devoted time to develop attic truss solutions resulting in their *Hatat system* of tribearing designs. Where a full height truss is prefabricated, roof members are kept to relatively small sections by using deep floor joist units built into the truss component. On truss shapes too tall to transport, the method illustrated in fig. 6.13 is employed. This of course is really a series of prefabricated components for site assembly using Hydro-Air's 'cam plates', these being prepunched metal plates for site nailing with special improved nails.

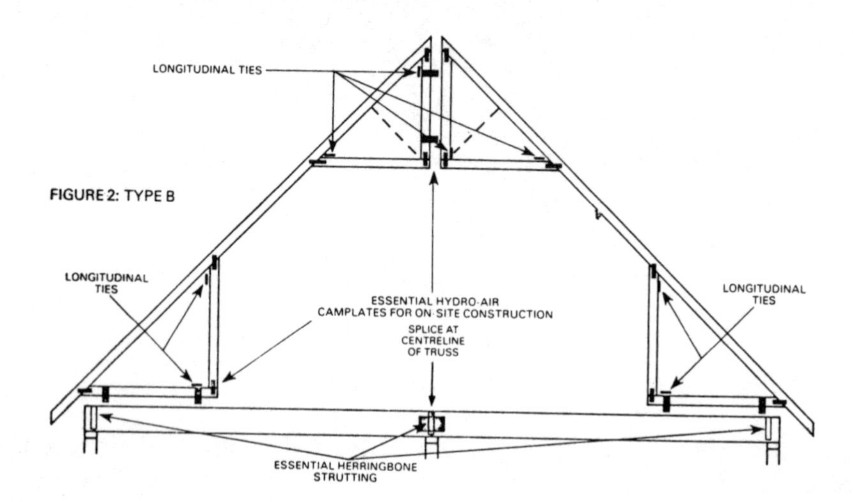

Fig. 6.13

Clear span attic trusses, i.e. not internally supported as the hatat system above, invariably use much larger timber sections of up to 47 mm × 225 mm. The availability of the attic may well be limited by the capacity of the fabricator – not the system owner – to avail himself of the larger sections in machined stress graded form. Large section stress graded timber is not generally held in stock by timber merchants nor indeed by those manufacturers who have their own stress grading equipment. It is, therefore, advisable when considering the use of attic trussed rafters to allow additional time for both their design and their fabrication, checking at the design stage with the fabricators on their ability to handle the design considered.

On attic trussed roofs it is particularly important to ensure that a good set of detailed site instruction drawings is prepared by the trussed rafter supplier, not just for the trussed rafters, but for the whole roof construction. Dormer windows, stairwells and room partitions within the attic, plus the complication of water tanks all necessitate careful design attention.

CHAPTER 7
Roof construction detailing

GENERAL

This chapter contains a number of construction details which relate to all types of roof construction and to avoid repetition in Chapters 2, 4 and 5, they have been grouped together in this reference chapter.

The items included are as follows:

STORAGE AND HANDLING OF TIMBER AND TIMBER COMPONENTS

Timber is one of the oldest building materials known to man and is still the least respected of all on most building sites. It is probably its resiliance to misuse which allows bad site practices in the storage and handling of timber to be tolerated. Generally it is only those timbers which are seen in the finished building which are afforded some respect and protection, whilst the often unseen structural timbers are frequently left unprotected and poorly stacked. Stress graded floor joists are not infrequently used as scaffold boards or barrow runs, and after having successfully survived these temporary functions are then built into the property.

The practicalities of construction work make it almost impossible to protect timbers completely, immediately after they are fixed on the house, but rapid enclosure of the roof structure and good ventilation will do much to prevent subsequent problems resulting from shrinkage on drying out.

Deliveries

It is not uncommon for the so-called carcassing timbers to arrive on site in a damp or even wet condition from the timber merchant for indeed most carcass timbers are not stored under cover. Such timber should be returned to the merchant for dry stock. If the builder has to accept wet timber, he must immediately take steps to achieve as much drying out as possible, before building the timber or components into his house.

Structural timbers such as rafters, purlins, ceiling ties, floor joists, etc., will often be delivered strapped with steel or nylon bands in house sets and may well be off-loaded by a lorry mounted crane or fork-lift truck. If the timber is at all damp then the straps should be split immediately, the timber inspected for quality and the consignment checked against the delivery note and the cutting schedule for the timbers required. Any shortages should then immediately be notified and rectified before construction starts.

The timber should be restacked on sets of bearers ideally 150 mm minimum off the ground and with thin sticks placed vertically in line with the bearers between each layer of timber to facilitate rapid drying. Any subsequent packs of timber placed on top of this stack should have their bearers again directly in line with the bearers of the pack below to avoid distortion of the timber, which could result in difficulties in fixing and certainly in poor quality in the completed building. Fig. 7.1 illustrates a poorly stacked set of timber.

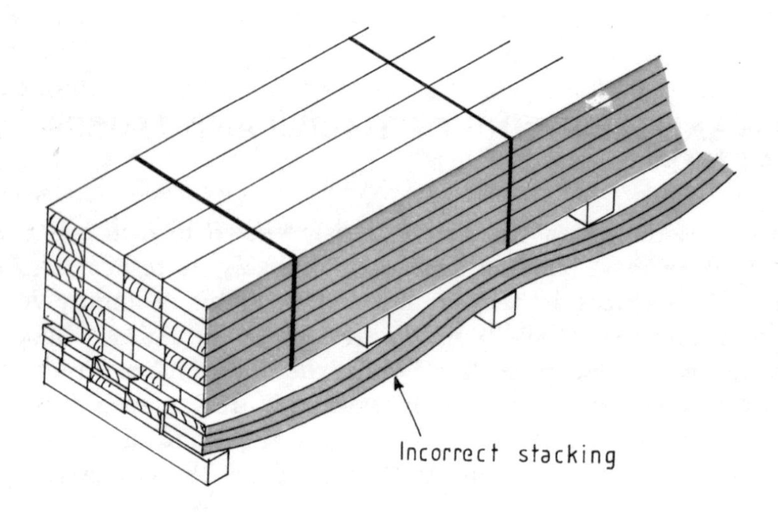

Fig. 7.1

Protection

Wet or dry timber should be covered loosely with waterproof sheeting, leaving the ends of the packs open for air to flow through the stack. This will avoid sweating within the pack and the possibility of resulting mould growth should the timbers be left in store for some time. If covered storage is available use it, even if moving it into store takes longer than leaving it where the lorry dropped it. Dry timbers are both lighter to handle and easier to work, and most importantly they present fewer maintainance problems in the completed building because of the reduced degree of shrinkage.

Manufactured structural timber components

Extra care is needed with the handling and storage of manufactured structural timber components. Rough handling can cause structural damage which may not become apparent until the component is built in and loaded. The cost of replacement at that stage can be very high, and of course can result in delays to the building process.

Long purlin beams of solid, laminated, or ply box or webb construction should be lifted from the delivery vehicle upright and not flat (see fig. 7.2). Webbing slings from a spreader beam should be used in conjunction with either a crane, fork truck or forks on a digger and then placed on level sets of bearers and protected as described above. Major structural items should be delivered to the site as close to the actual building-in requirement as possible, thus avoiding storage. Major purlin beams, for instance, should ideally be off-loaded from the delivery vehicle by crane and hoisted directly into position in one continuous operation. This of course avoids the costs of double handling and the problems associated with storage on a restricted building site. If storage is unavoidable, then prepare the area to receive the components *before* delivery, this will speed off-loading and again save significant costs.

The Building Research Establishment paper referred to in Chapter 5 (IP 14/83) makes reference to the unsatisfactory handling and storage of trussed rafters on most building sites. The International Truss Plate Association (ITPA) produce a bulletin on the subject, *Technical Bulletin* number three, which is available either directly from the ITPA or through the trussed rafter manufacturer. This gives guidance on both storage and handling of trussed rafters. The subject is also covered by BS 5268 part 3 in section 7, clauses 28 and 29.

Most trussed rafter manufacturers deliver the components on special vehicles on which the trusses are stacked vertically – they should remain that way on site during handling and, of course, in their working position. It is the only way in which they can be considered to be a self supporting component. Flat stacking, although sometimes unavoidable, is not in the writer's opinion considered satisfactory, and certainly requires far more attention to correct preparation of the storage area on site.

Fig. 7.2

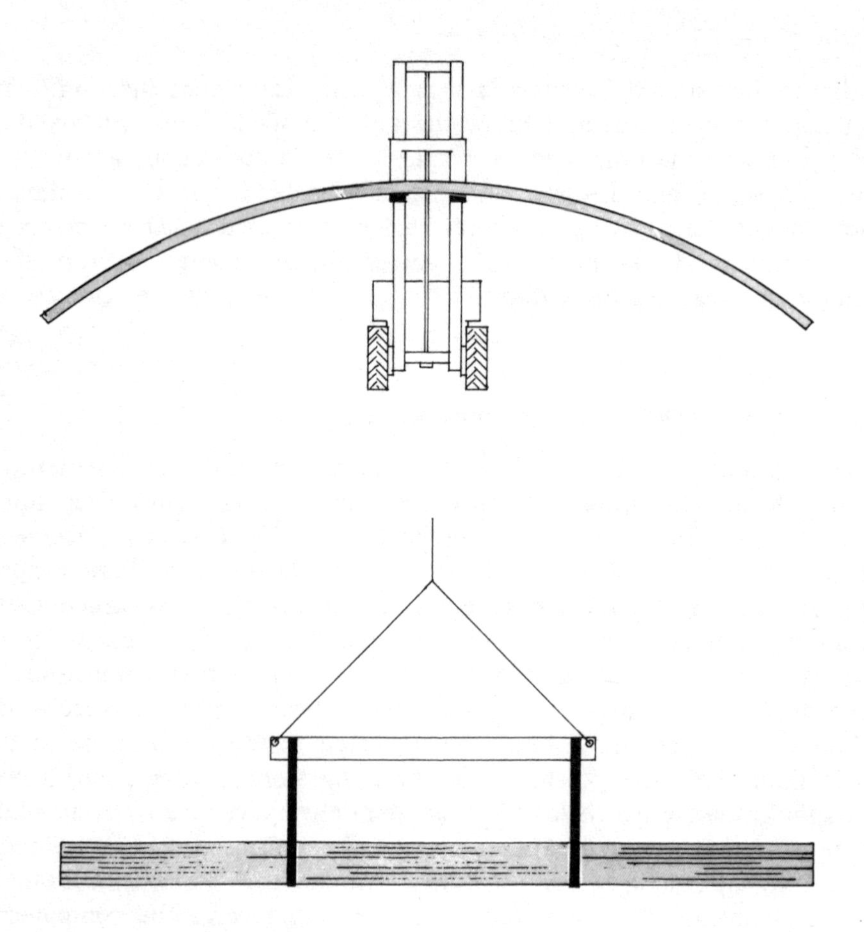

Preparation of the store area

Trussed rafter manufacturers invariably deliver the components to an agreed time schedule to the builder. For this reason there is no excuse for the site staff not being ready to receive the components. Preparation of the storage area, particularly on larger building sites, should be carried out well before delivery, and on the very largest sites a permanent truss storage system built as part of the site set-up (see fig. 7.3).

A simple scaffolding frame is all that is required to support this relatively light component, with main bearers placed at approximately wall plate position and possibly further bearers placed at approximately bottom chord third point positions. It is fully appreciated that more than one truss span is required on any one individual site, but careful consideration of the plans will give a satisfactory set of support positions. Having set out the horizontal supports, which should be at least 300 mm off the ground, a triangulated vertical support system should be placed at the back of the intended stack, giving support to the trussed rafters at approximately their quarter span points. This back support should be inclined backwards away from the horizontal bearers to prevent individual trusses falling forward causing damage.

The delivery

On arrival of the delivery vehicle check the trussed rafters against the delivery ticket and the materials requisition schedules or plans before off-loading the lorry. Assuming that all is satisfactory off-loading can commence, giving the trussed rafters a physical inspection as unloading proceeds.

Many trussed rafter manufacturers strap their trussed rafters in packs of ten or twelve and if mechanical handling is used on site they are best left in their strapped packs at least until placed in the storage rack. A spreader beam as described above should be used with webbing slings passed through the quarter points on the top chord or at some similar node point. The pack should be stabilised whilst lifting, either by direct hand control if at ground level or, if the pack is to be lifted straight onto the roof plate, then by ropes attached to the heel joint positions. Strapped packs should then have the bands cut whilst they are in store on site, because the timber used in trussed rafter construction is invariably quite dry and any moisture uptake on site, whether the stack is well protected or not, will cause swelling of the timber resulting in the bands crushing those trussed rafters on the outside of the pack. Where mechanical handling is not available then obviously the straps must be cut on the lorry and the trusses manhandled on to their storage racks.

Again it must be emphasised that the trussed rafters must be carried *vertically* not horizontally. Horizontal transportation, frequently with one man at a position about one quarter way up the rafter, will result in serious 'whipping' of the trussed rafter and can cause plate disturbance and therefore damage to the component. Some trussed rafter manufacturers occasionally deliver the trussed rafter inverted for transport economy reasons, these should be off-loaded in their inverted position, carefully laid on the ground and then the peak lifted into a vertical position whilst the bottom chord is still on the ground. The truss should not be flipped over whilst being supported at the heel joints.

Figs. 7.3-7.5 inclusive illustrate some of the points discussed above.

Protection

When the delivery has been safely stored in the racks, plastic sheeting protection should be fitted over the top chords at least, and loosely over the ends of the package. Care must be taken to maintain adequate through flow of air to prevent sweating within the stack.

Once the trussed rafters are constructed into their roof form, enclosure of that roof should be as fast as possible. At least felt and battens should be applied to prevent unnecessary wetting of the trussed rafters and, of course, the building structure below.

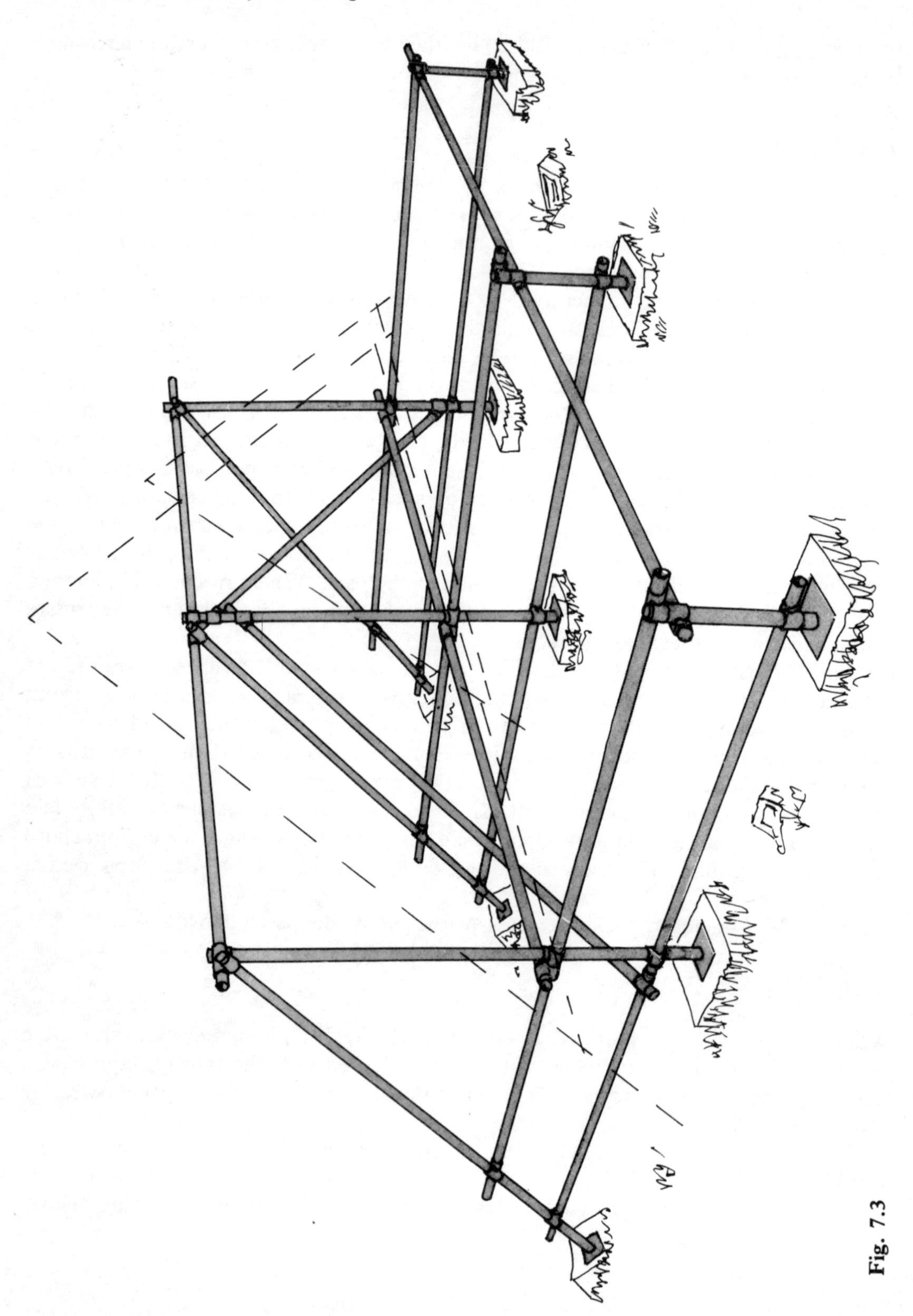

Fig. 7.3

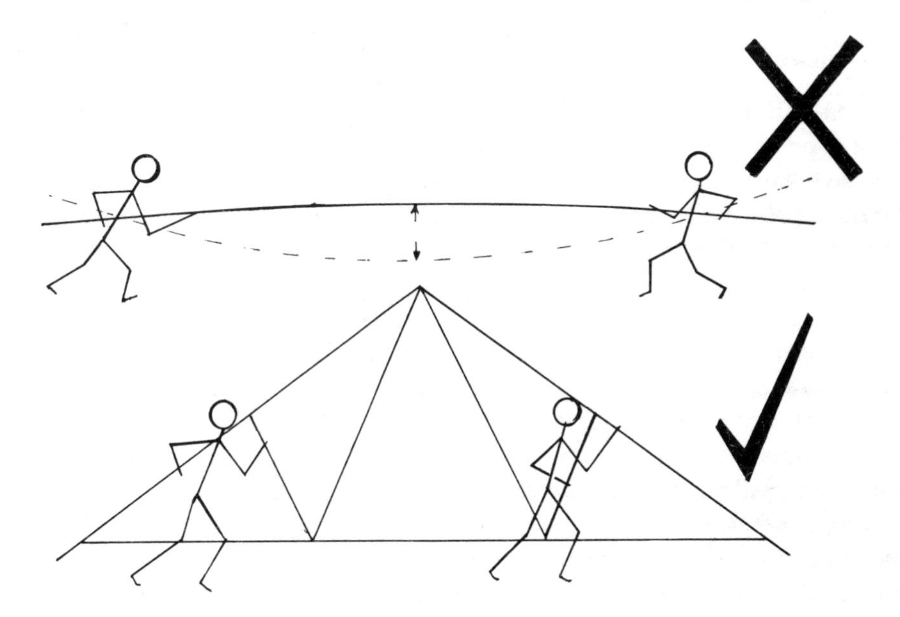

Fig. 7.4

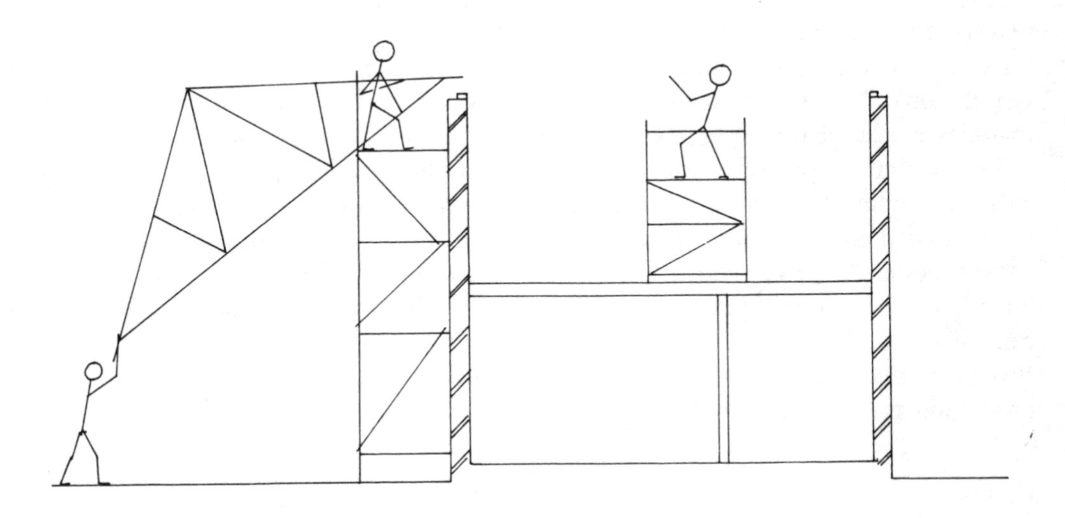

Fig. 7.5

Man-power

Trussed rafters are very light components and for this reason are generally easily handled by two men. Roof spans above 9 m and pitches above 35°, however, result in a very awkward frame to handle and additional assistance will be required. British Standard 5268 part 3 clearly states that trussed rafters forming girder trusses should be fabricated into their girder at the factory. This is not common practice at present but will be so in future and handling of these obviously more heavy components must be considered before their arrival on site. Man-handling will, of course, be possible but significant numbers of men will be required, particularly where a large span triple truss girder is required.

The attic truss form of course presents its own problems, being generally of much larger timber section and by its very nature invariably of a steep pitch. These components will be much heavier and require mechanical handling or up to six men to off-load safely, transport them and stack them on site. It is strongly recommended that attic trussed rafter forms be treated in the same way that precast concrete floors would be treated, that is carefully timed deliveries in conjunction with mechanical handling on site, taking the component direct from delivery vehicle on to the prepared building structure.

PRESERVATIVE TREATMENT

Timber maintained at a moisture content below 20% is unlikely to be attacked by timber decaying fungii. However, the same does not apply to insect attack. Building legislation does not at present require structural timbers to be preservative treated (except those in flat roof constructions), with the exception of an area of England affected by the house long horn beetle. This area is defined in the Building Regulations. NHBC require external claddings to be treated with preservatives, this includes fascias, barge boards and timber soffits.

The various types of preservative and their application processes are set out in British Standard 5268 part 5 'Preservative Treatments for Constructional Timber'. It is not intended to enter here the detail of timber treatments, suffice it to say that two basic types are used for constructional timbers in this country, the first being a water-borne preservative system, the second being organic solvent-borne. Their application can be by total immersion or vacuum and/or pressure depending on the type used. Brushing of these preservatives is generally not adequate to afford a high degree of protection for structural work.

Costs

The cost of timber treatment when taken in context with the overall cost of building a house is minimal, and it is therefore to be strongly recommended whether required by the regulations or not. For the traditional cut roof or those timbers site fitted to the other types of roof discussed in this book, the additional cost is approximately 20% of that of the raw material. With trussed rafters the additional cost varies between approximately 5 and 10%, depending upon the size and complexity of the trussed rafter in question.

With the increased emphasis on high insulation values, the roof structure is becoming increasingly cold. Although the roof is now usually well ventilated the cold metal fixings, be they nail plates, truss clips or gable straps, can attract condensation which could clearly be absorbed by the timber to which they are fixed. Preservative treated timber will obviously resist any degradation which could occur. Ventilation has of course slightly increased the risk of insect attack in roof spaces, although incidence of this is extremely low. Experience with well ventilated roofs is relatively limited, however, the regulations requiring ventilation having been introduced only in the past few years. For this reason again it seems prudent to preserve the roof structure.

A question which has caused some concern in recent years is the possible corrosion of the galvanised punched metal plates used on trussed rafters, when they are used in conjunction with the water-borne 'CCA' type preservatives. The concern was that the chemicals used in the treatment could cause the zinc in the galvanising to be oxidised, thus leaving the mild steel unprotected with the resulting risk of rusting. The Building Research Establishment information paper IP 14/83, referred to earlier, contains comment on a corrosion survey carried out on trussed rafter roofs. This states 'Thus whilst CCA does lead to more corrosion, this is still at a low level and on present evidence is unlikely to be of any structural significance'.

Particular care is needed with CCA preservatives to ensure that the timber is dried out thoroughly before the metal plates are fixed at the factory. (BS 5268 part 3 clause 11 deals with preservative treatment of trussed rafters.) This significantly reduces the risk of plate corrosion by the preservative. It must be said, however, that the practicalities of carrying out this drying at the factory, bearing in mind the relatively short notice given to the manufacturer to produce the trusses, is impractical. It is therefore strongly recommended that CCA preservatives are not used with trussed rafter assemblies, the specification should therefore call for an organic solvent based preservative.

In situations of high hazard in the roof space, e.g. over a swimming pool, then preservation should be considered essential. Trussed rafters used in these conditions should use stainless steel plates, and either type of treatment system would then be acceptable. Other conditions of high hazard at, in the writer's opinion, those areas of the country immediately adjacent to coastlines, where the corrosive salt-laden atmosphere is of course encouraged to enter the building via the ventilating system.

Preservative identification

When preservation has been specified, there is obviously a responsibility on the part of the contractor to make sure that the specification has been met. Many of the organic solvent based preservatives carry a faint tracer dye (often light reddish brown) but some manufacturers use a clear preservative. Organic solvents fresh from the factory will carry a distinctive oily smell and this clearly would indicate that the preservative has been used, but can in no way identify the application process. Simple preservative testing kits are available from most of the preservative manufacturers, if any doubt exists. For security it is best to specify that the trussed rafter manufacturer provides a signed certificate of treatment for the trusses in any one consignment.

The CCA preservatives, being water-borne, have little or no odour. Identification is generally easily made by the light green colour imparted to the timber by the chemicals involved. If any doubt exists, simple tests can be carried out, details of which are available from the manufacturers of the treatment system or from the treatment plant used. Again, for security, a certificate of treatment should be obtained.

WALL PLATES AND FIXINGS

The wall plate is the foundation for the roof. Care in setting out and bedding the plate will not only make the roof construction easier, but will result in a more sound construction. Poor alignment in particular with the purlin in a traditional roof will mean that each common rafter has to be individually fitted, making precutting of the birdsmouthing to a master pattern impossible. The additional time in cost can be considerable and again can cause delays to the construction process. Fig. 7.6 illustrates the effect on a rafter of poor alignment.

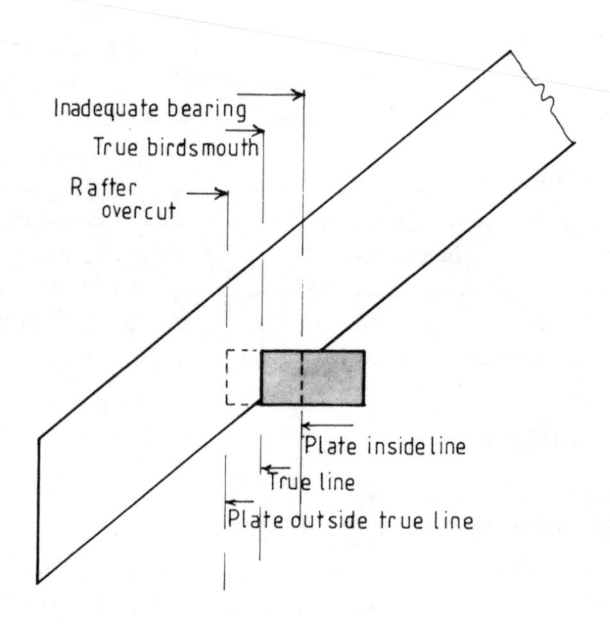

Fig. 7.6

Wall plates should not be less than 38 mm thick × 75 mm wide, 50 mm × 100 mm being a common section used in house roof construction. The wall plate width for trussed rafters should not be less than 75 mm, or 0.008 × the span of the trussed rafters. Half lapped joints should be made at all intersections and well nailed together (see fig. 7.7).

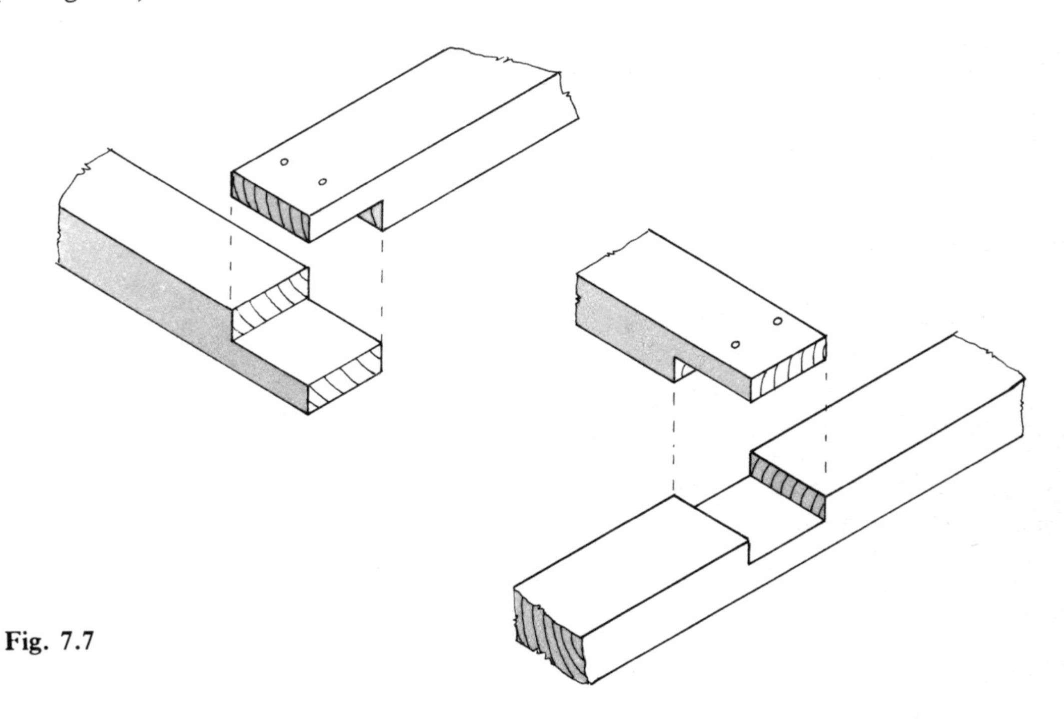

Fig. 7.7

Strapping

Where the roof is in an exposed location, or the tiles or roof covering are very light in weight, the roof designer may require the wall plate to be strapped down to the supporting walls below. The straps used are standard components readily available from builders merchants. Two main types of straps exist, firstly that built-in to the course of inner skin blockwork and secondly that nailed to the inner skin blockwork. Both are securely fixed to the plate usually by nailing (see fig. 7.8).

On timber framed housing the plate may be adequately secured to the frame of the house by nailing at the centres and to the specification laid down in the nailing schedule provided by the house designer. If the timber frame method uses a separate binder or wall plate across the top of the panels, this needs to be half lapped in relation to the panels to which it is fixed. This detail is shown in fig. 7.9.

Fixing the roof to the plate

Having secured the plate, the timbers, trusses or trussed rafters need securing to the wall plate itself. This can be done by skew nailing on all forms of roof, but on trusses

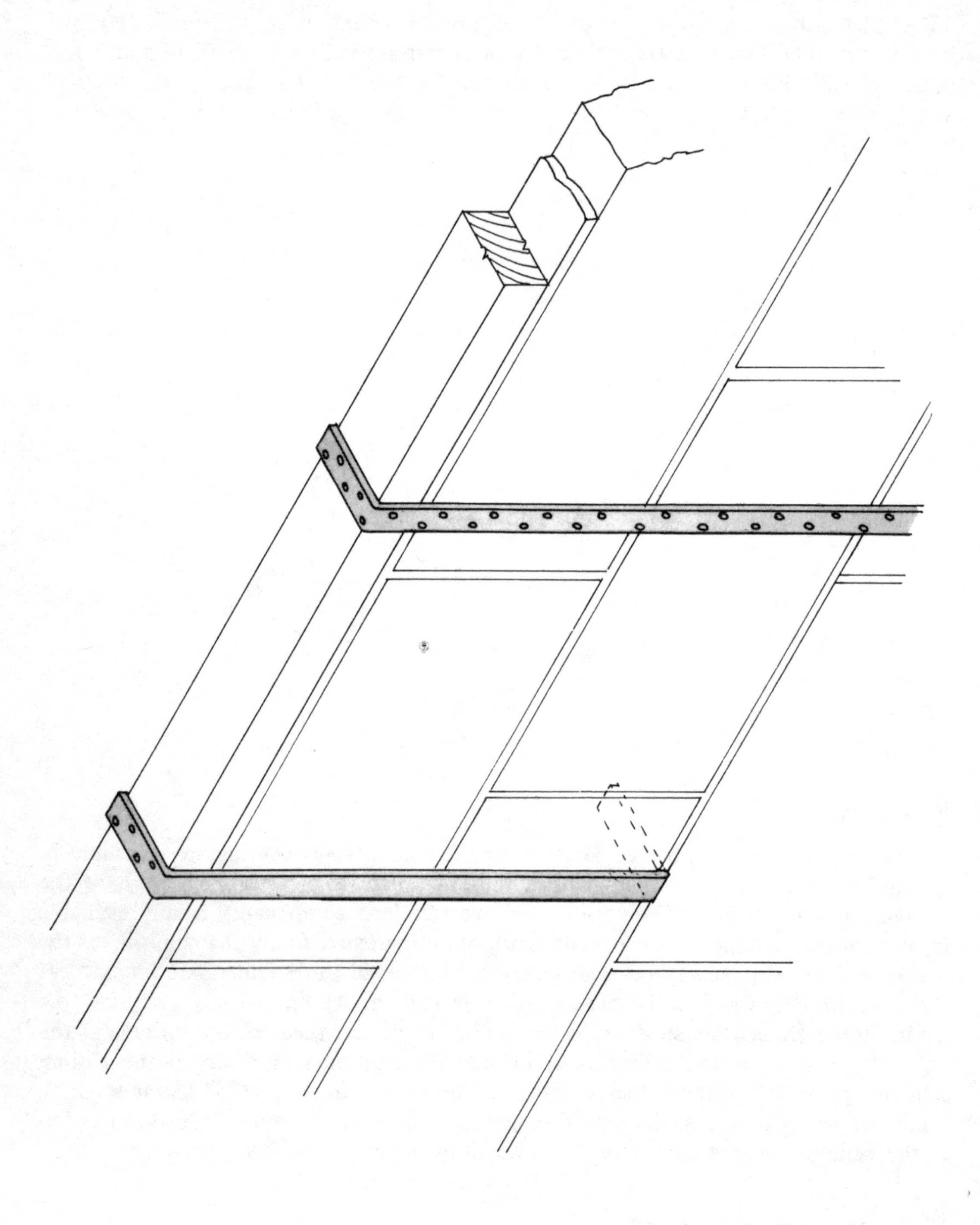

Fig. 7.8

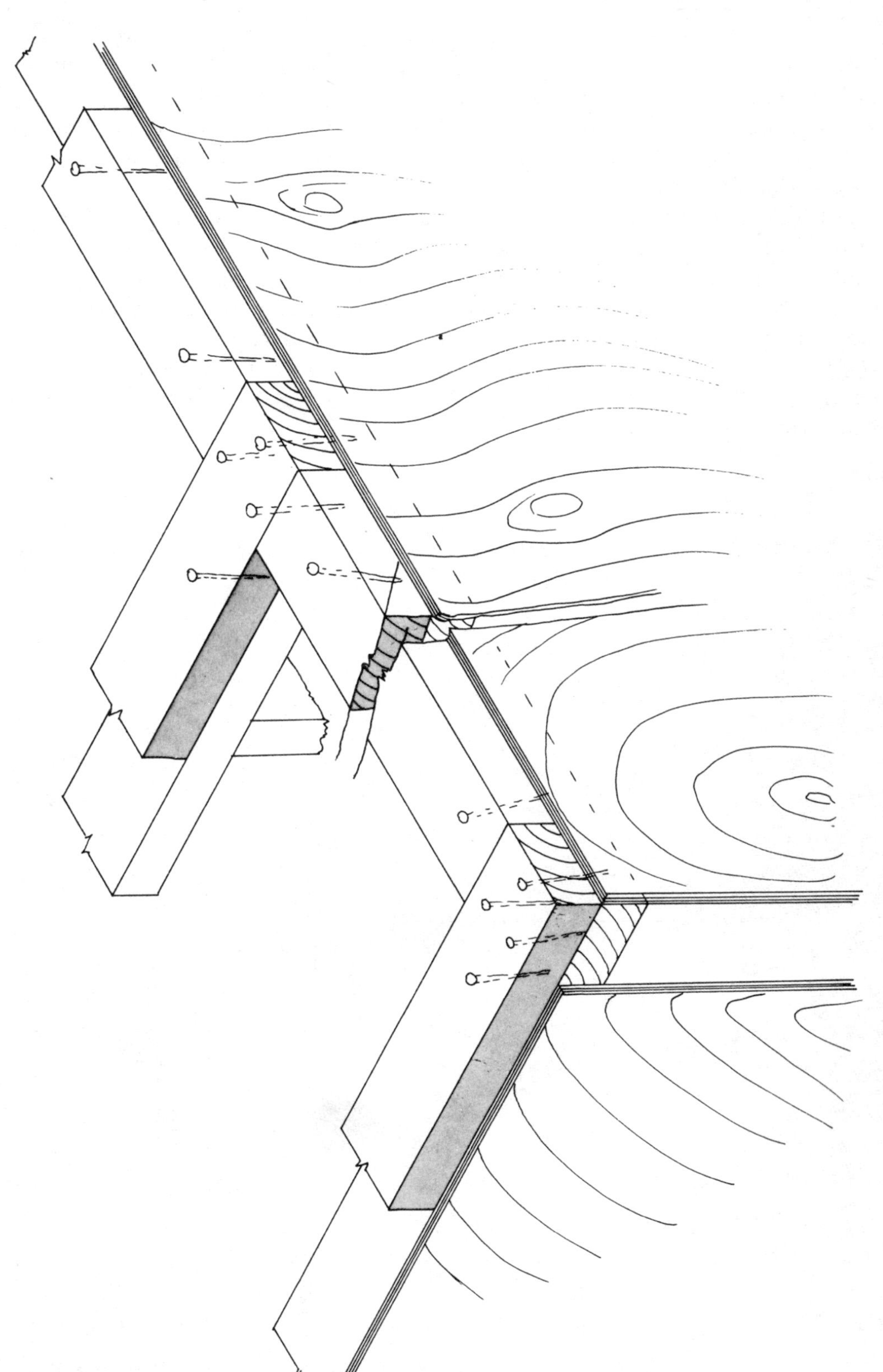

Fig. 7.9

or trussed rafters there is a danger of the skew nails disturbing the joint, or themselves being deflected by the plates or connectors, resulting in an ineffective joint. Truss clips or framing anchors should be used as illustrated in fig. 7.10, and these again are readily available items from the builders merchant. Alternatively, and particularly on the bolt and connector roof, the truss may be strapped directly to the wall below, as indicated in fig. 7.11.

It is *strongly recommended* that for trussed rafter roofs the truss clip is used, whether or not it is specified on the drawing. Skew nailing through the connector plate at the heel invariably results in splitting of the timber in the bottom chord, simply because the nail is placed so close to the cut end of that particular member. Unless the nail is driven at an angle which would result in it emerging from the bottom of the bottom chord, it can also disturb the penetration of the teeth on the opposite side of the heel joint, thus again rendering one of the most heavily loaded joints in a trussed rafter, weakened.

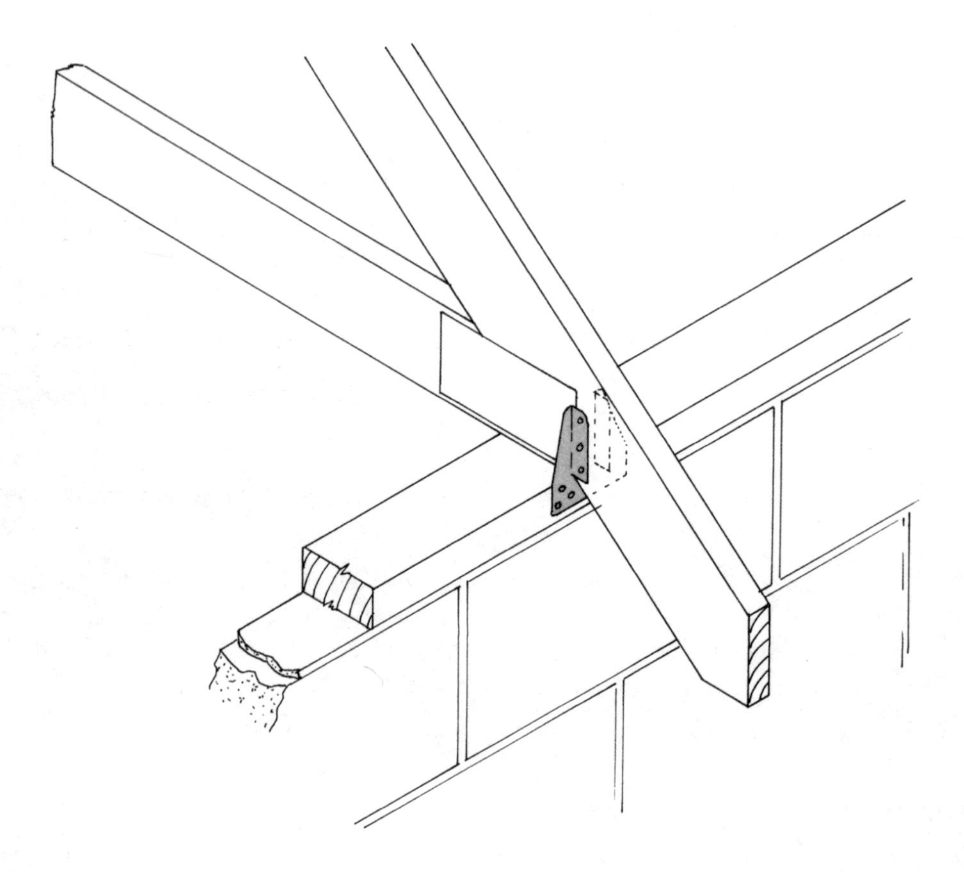

Fig. 7.10

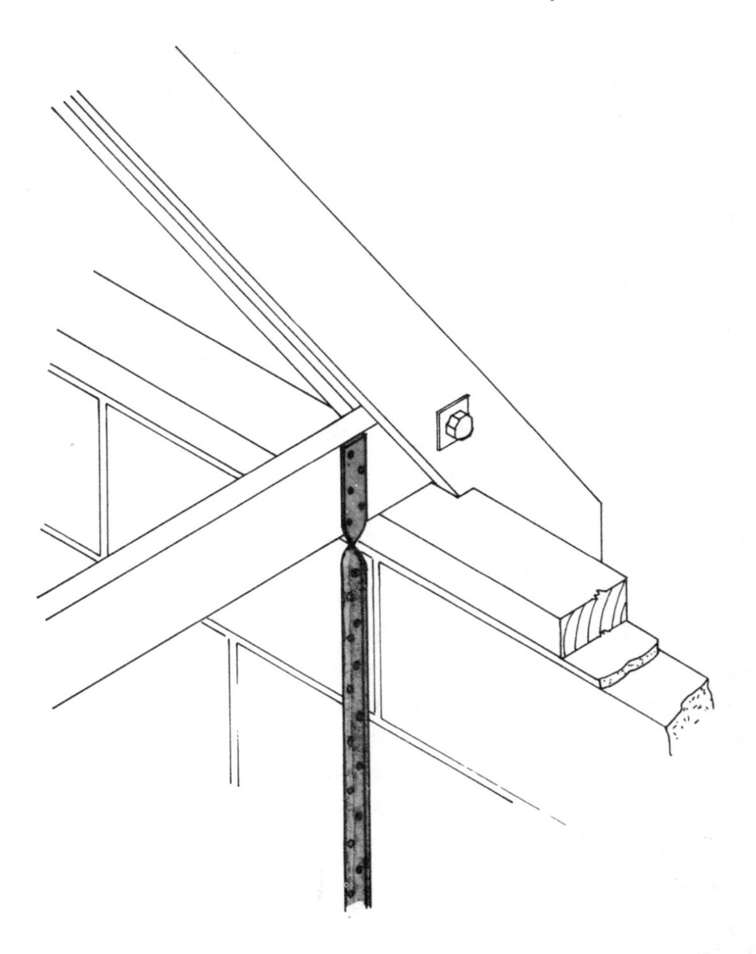

Fig. 7.11

GABLE ENDS, LADDERS, GABLE RESTRAINTS AND SEPARATING WALLS

Gable ends

There are three principle alternatives for the detailing of roofs at the gable ends or 'verge':

(1) To provide an overhang similar to that at the eaves complete with timber trim or barge boards and/or soffit
(2) To simply bed the tiles onto the top of the wall, usually on a strip of durable material to support a minimal overhang to the tiles
(3) To carry the gable wall up past the roof to form a parapet

Detail one

The detail chosen by the architect will depend upon the building style rather than any structural considerations. Taking the first option above requires more attention to the structure because of the overhang. The provision of the supporting overhang will depend upon the roof construction used. Where a purlin is used to support the commons this can be taken through the gable walls and projected to give the overhang required, thus supporting the rafter at that point. Similarly, the ridge may be projected to support the common rafter at the top. Occasionally the wall plate itself may also be continued through the wall to provide support at the foot of the common rafter. If this is not done then some noggins will need to be built into the wall and fixed to the last common rafter on the roof proper, cantilevering out to carry the rafter of the projecting roof.

The barge board will be fitted to the last rafter, its top being level with the top of the common rafter. In the construction of some older houses the barge board was deep enough to project above the top of the rafter to a height equal to the thickness of the battens and the tiles, and then a capping timber was nailed to the top of the rafter to complete the weatherproofing. This capping is very vulnerable and of course is a significant maintenance problem, and for this reason the detail is seldom used today. Where exposed rafter feet are used it is not uncommon to leave the purlin, the ridge and any other supporting members (projecting out to carry the gable overhang) exposed to view. More frequently, however, the underside of the overhang is closed off with a soffit to match that of the main building. Where a soffit is required the soffit board can be fixed to the underside of the projected rafter, and on to battens fixed to the gable wall. The above detail is suitable with either the traditionally cut roof or the bolt and connector roof, both having substantial purlins, ridges and, if necessary, large wall plates.

The gable ladder

On trussed rafter roofs no purlin or ridge board exists, therefore alternative means of supporting the overhang is required. The 'gable ladder' is used to provide the roof support (see fig. 7.12).

The gable ladder can be constructed on site, but is more usually provided either assembled or as a set of precut components by the trussed rafter manufacturer. It is common practice to use the same timbers in framing the gable ladder as are used in the trussed rafters themselves, thus ensuring good alignment of the roof. The gable ladder is a simple nailed assembly, itself nailed through one of the gable ladder rafters directly to the last trussed rafter on the main roof. The brickwork is then built around the 'rungs' of the ladder to fix it securely in place. Significant overhangs beyond the gable end can be achieved using this detail, but beyond about 450 mm particular care must be taken in fixing the gable ladder, ensuring that it is correctly designed to carry not only the load of the roof but also the operative working on the roof. Barge boards can be fixed as previously described and, with the fascia continuing through from the main roof, the typical barge board to fascia detail can be achieved. Wind uplift on very wide verge overhangs may require the gable ladder to be strapped down to the wall.

On timber framed housing it is quite common practice to use a completely prefabricated verge unit, this comprising the gable ladder, prefixed soffit, and prefixed

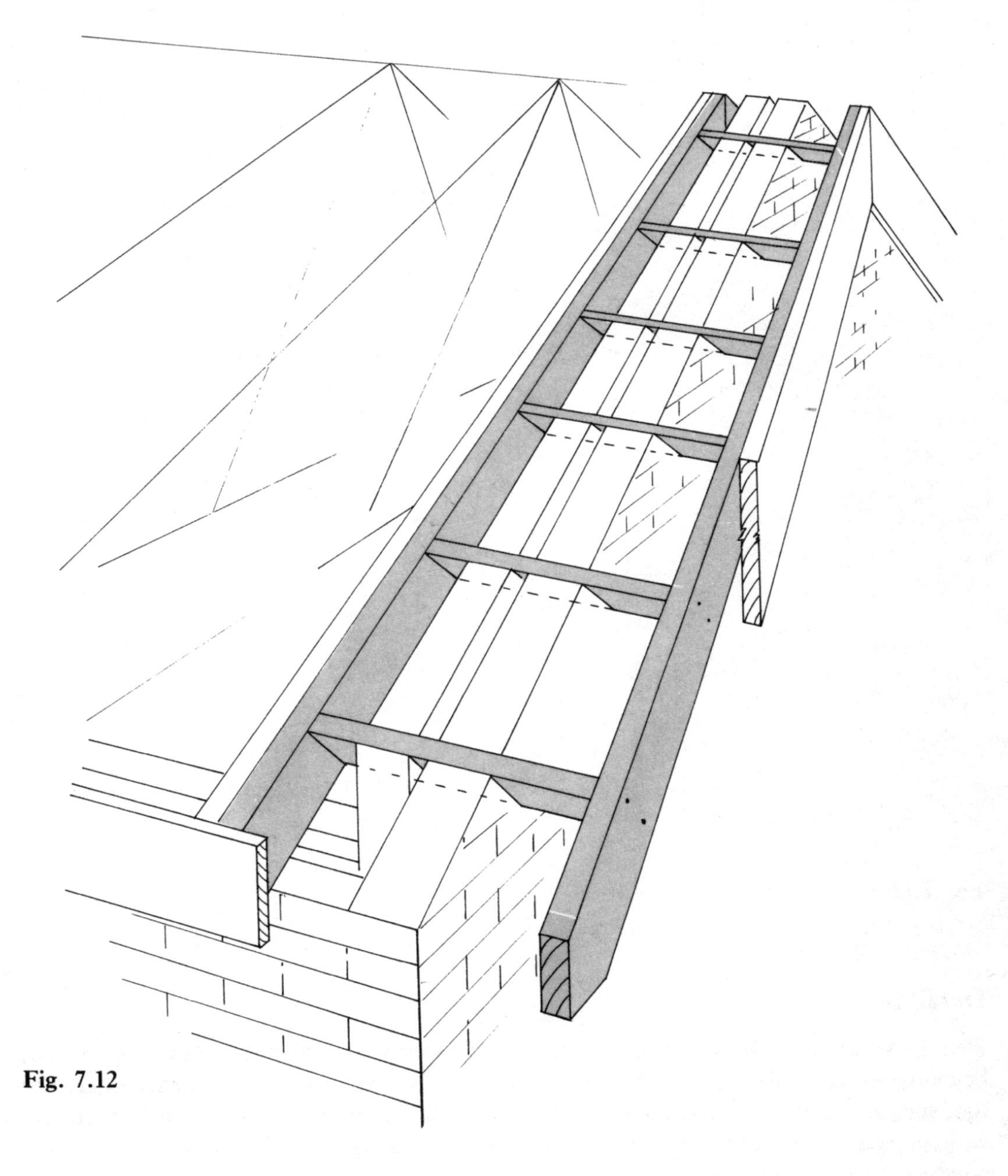

Fig. 7.12

barge boards. This unit will be nailed, as illustrated in fig. 7.13, to the last trussed rafter of the timber framed house and, supported on the timber framed gable end panel, it will cantilever over to give the desired gable end overhang. Care must be taken to ensure a settlement gap between the brickwork skin and the soffit to allow the timber frame to settle independently of the brickwork without disturbing the true roof line. The gap should be filled with a compressible filler.

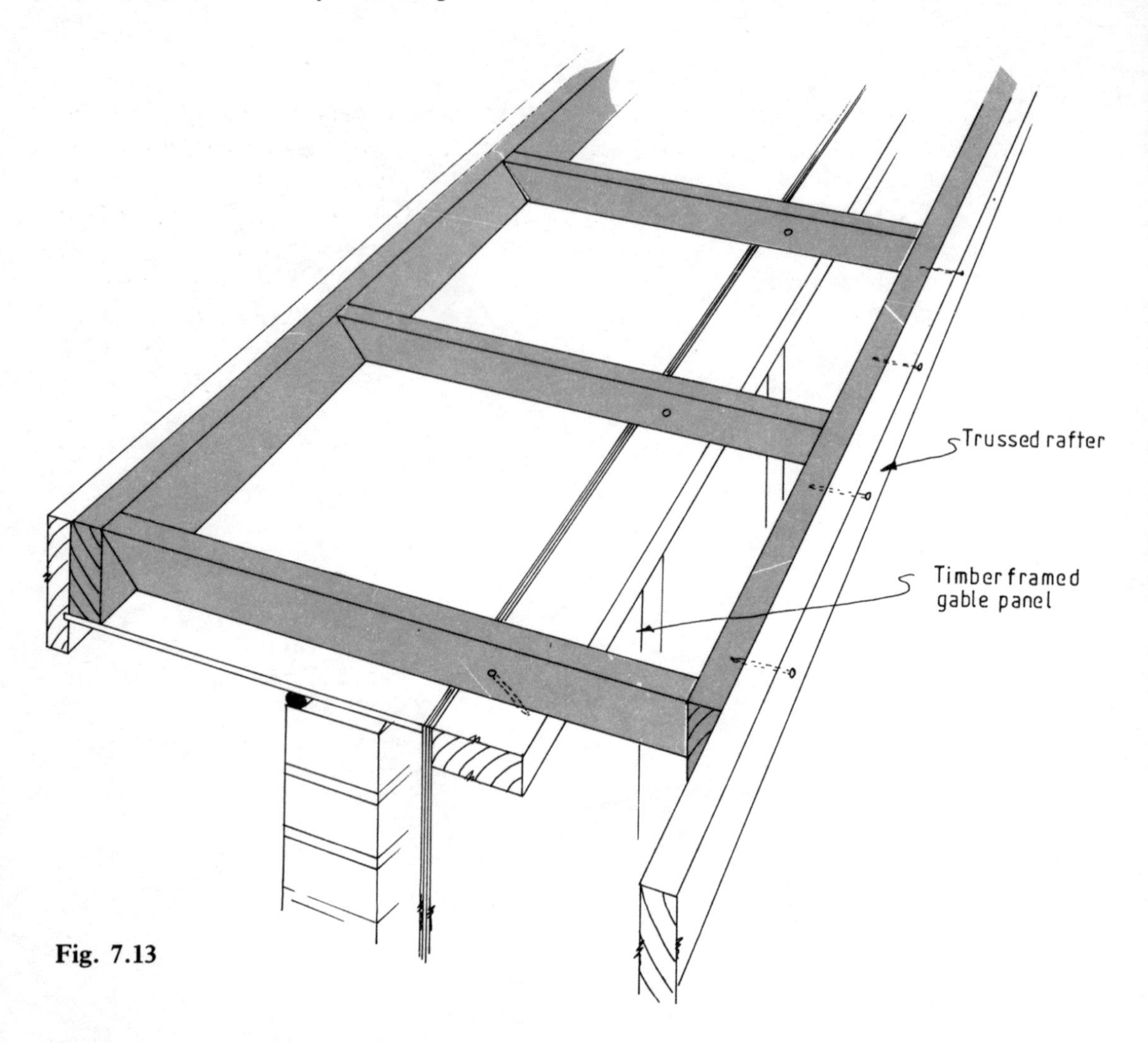

Fig. 7.13

Detail two

Detail two above requires little description. There is no timber structure to the roof beyond the last common rafter on the inside of the gable end. As stated earlier, the roof tiles are simply bedded on to the brickwork gable end giving a minimal overhang, with gaps between the tiles and brickwork and the laps of the tiles simply being pointed with cement mortar.

Most of the principle tile manufacturers now have what they term to be a 'dry verge' system. These systems use plastic extruded units to form a simple but neat and weatherproof gable end trim where no barge board is required.

A further option available from the tile manufacturers is to use a special verge tile, which in visual effect folds the tiles over the end of the roof down onto the brickwork again giving a neat weatherproof and maintainance-free finish. No timber work is involved in either of the proprietary systems.

Detail three

The parapet wall presents its own problems of weathering between the wall and the roof abutment. There are no structural problems as far as the timber roof structure is concerned, the last common rafter or trussed rafter simply being placed close to the inner skin of the gable end wall. The trussed rafter in particular should not be fixed to the wall, thus allowing natural deflection to occur in the trussed rafter roof independent of the brickwork. The weathering between wall and roof is generally achieved by means of lead soakers placed between the tiles and turned up the face of the abutting wall, these being covered in turn by stepped lead flashings fitted in to the brick courses.

The detail at the abutment will be influenced by the shape and style of tile used for the main roof. Fig. 7.14 shows a stepped lead cover flashing frequently used with heavily rolled tiles and pantiles.

Fig. 7.14

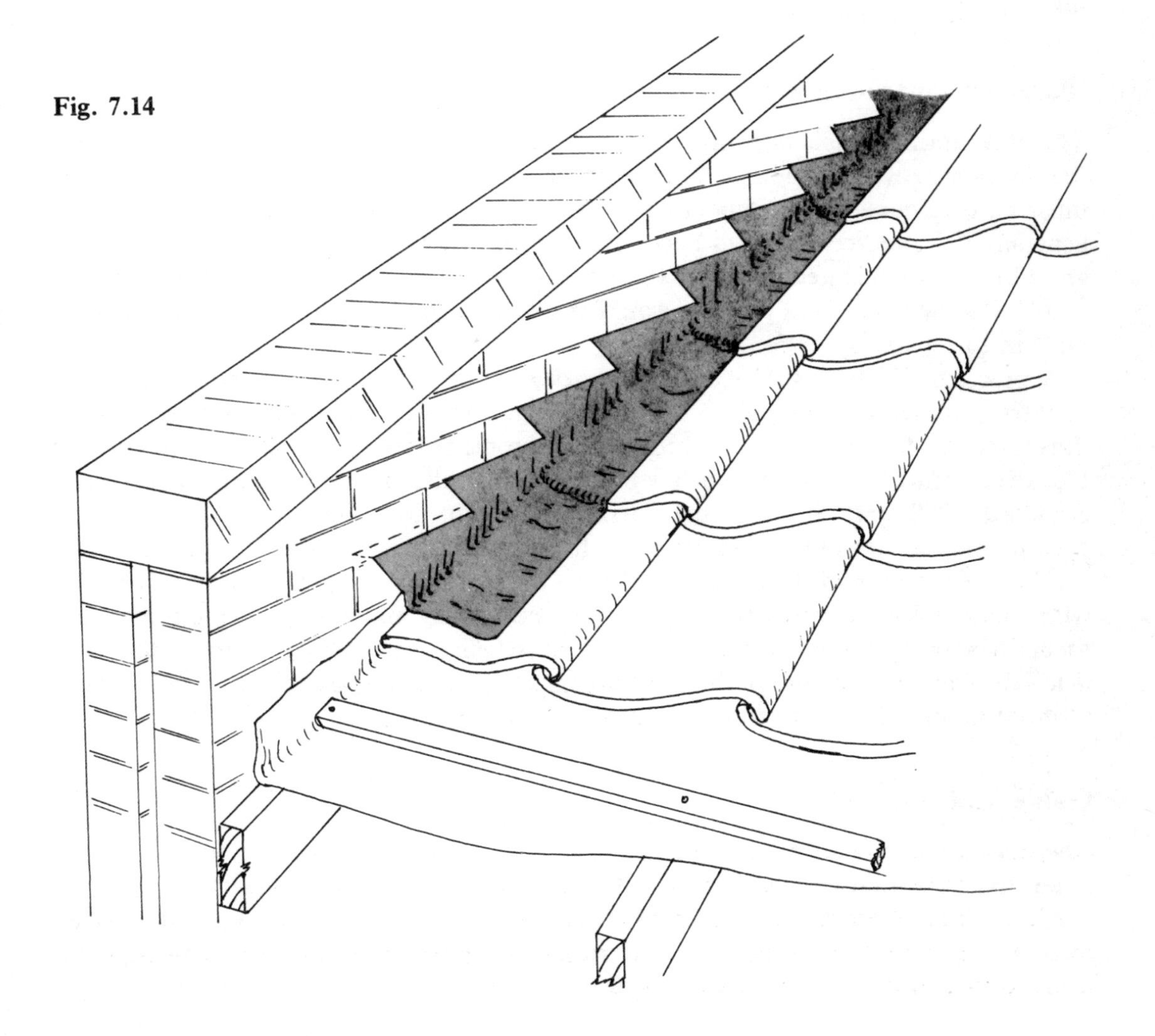

An alternative and particularly neat detail with the flatter type of tile is to use a very small secret gutter between the roof tiles and the parapet wall. The tiles are simply stopped some 25-38 mm short of the parapet wall, with a continuous length of lead dressed underneath the tiles with a welt or water check, down on to a small supporting batten between the last rafter and the wall, and then up the face of the wall to a height approximately matching that of the tiles themselves. Dressed over the top of that is the stepped apron flashing in the normal manner.

Proprietary roof tiling systems

It is not intended to discuss here the merits of the various roofing systems offered by the tile manufacturers. This is considered outside the scope of this book, which concentrates on the timber roof structure rather than the coverings. It is useful, however, to be aware of the many roofing details produced in the various manufacturers' tiling manuals of technical literature.

'Party' or separating walls

The other form of gable occuring in roof structures is that unseen gable constructed at the division between dwellings in pairs or terraces of houses. The wall in that position must be continuous to the underside of the tiles to ensure adequate fire break, and generally it is not acceptable to have timber members built into this wall, unless they are adequately separated between the houses with a material which will impart one hour fire resistance. Good practice would therefore dictate that timber in general is not built in to separating walls.

A problem which occurred in the early days of truss and trussed rafter use in this country was that of 'hogging' over these separating walls between buildings; this is illustrated in fig. 7.15. To overcome this problem, the party wall brickwork or blockwork must be kept down below the top of the rafter line by some 25 mm. British Standard 5268 part 3 limits deflection to 12 mm for roofs up to 12 m span, consequently at maximum deflection of the trussed rafter roof there is still a 12 mm gap between the top of the rafter and the top of the brickwork. This gap must be filled with compressible yet non-combustible material (usually mineral wool), the tile battens themselves being the only timber item to pass from one building to the other. The detail described avoids the problem of hogging and its unsightly effect on the roofs of terraced houses. Fig. 7.16 illustrates the correct detail.

Gable wall restraint

The question of pressure on one end of the gable end of a house and suction on the other end was discussed in chapter five. Fig. 7.17 illustrates this effect.

Above the wall plate the gable end is of course free standing brickwork and on steep pitched roofs this area of gable brickwork can be quite large, and the resulting wind loads quite significant. The Building Regulations require these external walls to be

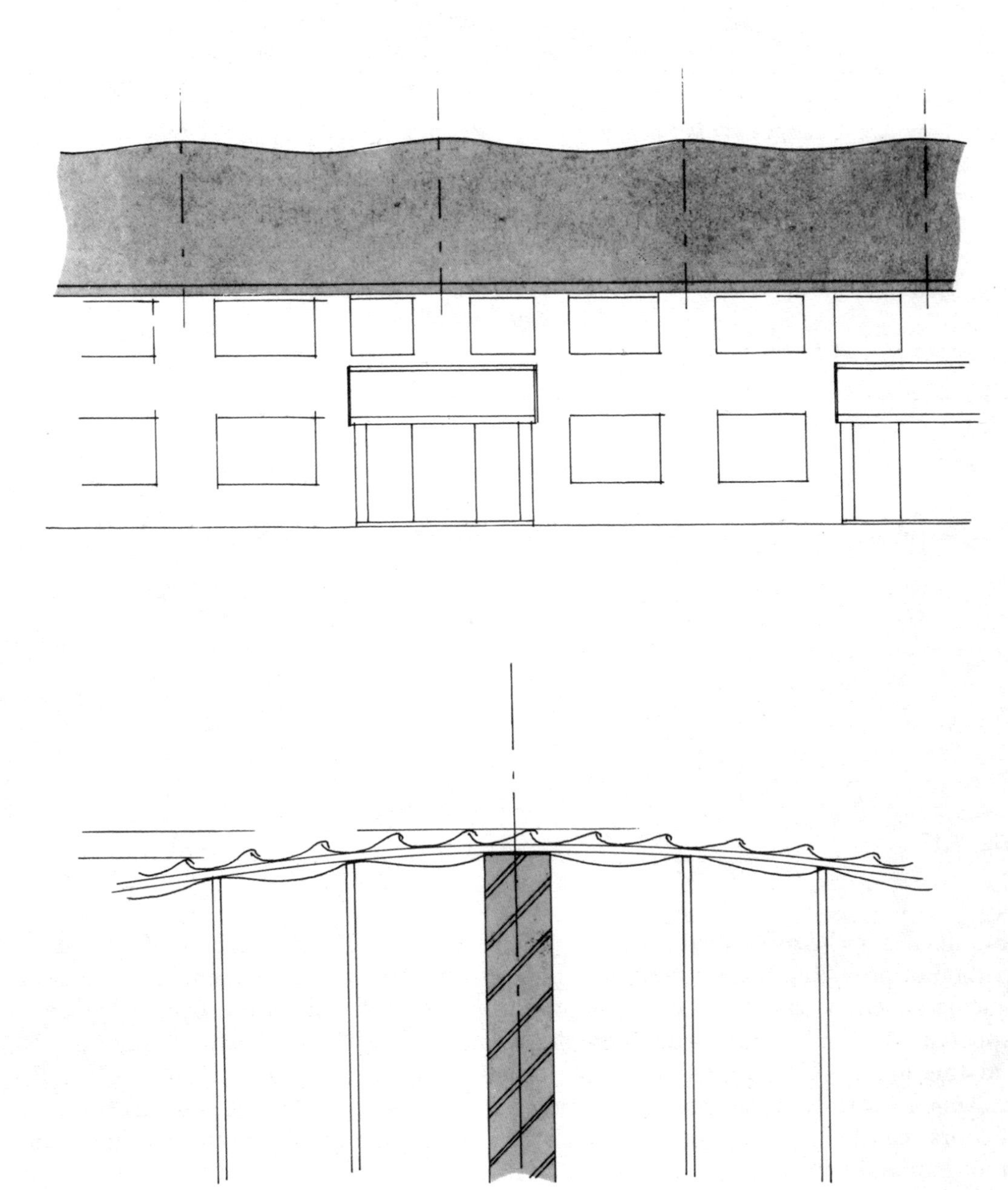

Fig. 7.15

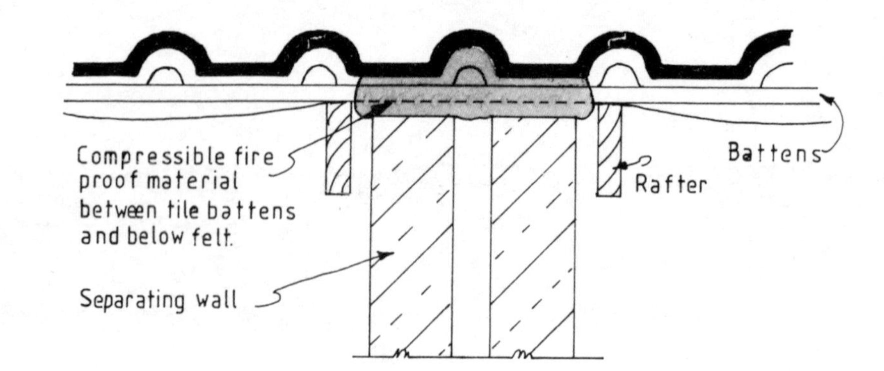

Fig. 7.16

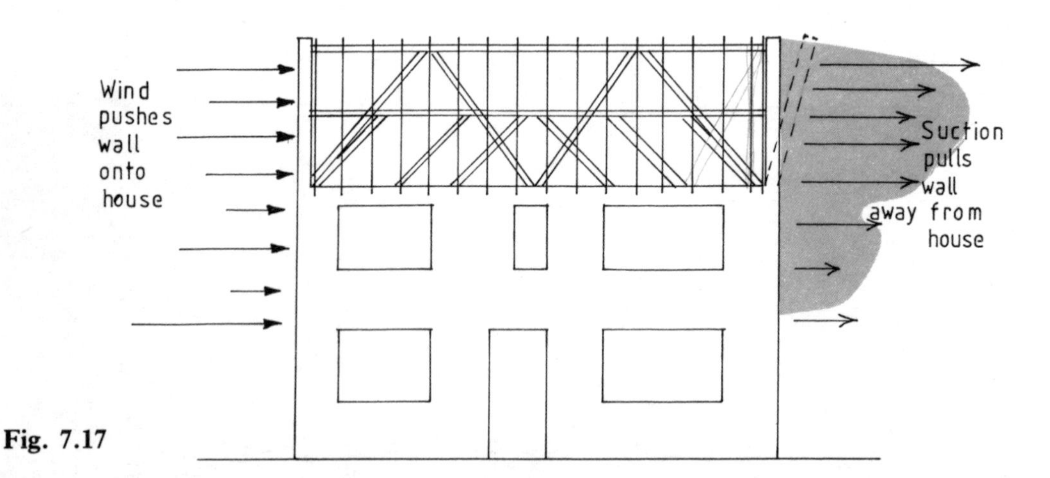

Fig. 7.17

restrained and British Standard 5268 part 3 appendix B gives details of a suitable method of providing the restraint on buildings up to three storeys high, and with roof bracing generally as described in Chapter 5 of this book and in the British Standard Appendix A. Similar bracing and restraint will apply to all roofs, however constructed, and this must not be forgotten simply because the details which are now generally available invariably show only trussed rafter roofs. With a traditional roof the restraint of course can be provided more simply by ensuring good fixity for purlins, ridge, and other longitudinal members into the gable ends.

Restraint is generally provided in trussed rafter roofs by fixing galvanised steel straps across at least two trussed rafters into the gable end. These contain the suction loads holding the brickwork into the roof, whilst blocking pieces fixed between the wall and the first trussed rafter and between the first trussed rafter and the second trussed rafter guard against the gable being blown into the roof. The ties should be of 30 mm × 5 mm minimum in cross-section and are generally prepunched with nail holes to facilitate ease of fixing on site. Fig. 7.18 illustrates a typical set of gable restraints.

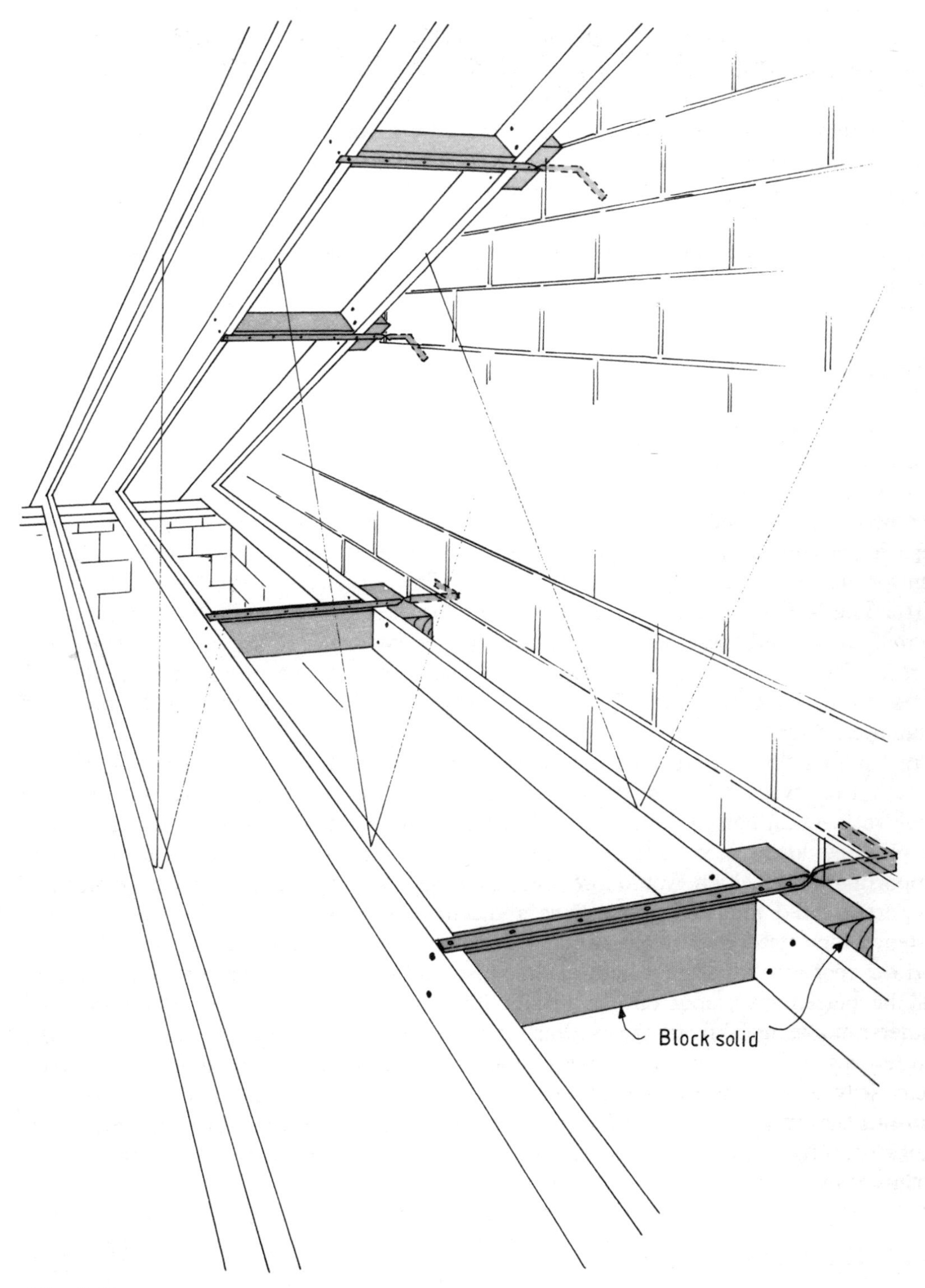

Fig. 7.18

It can be seen that these straps occur at both ceiling and rafter on gable ends, but on the ceiling they tie only on separating walls where they pass through them to form a continuous link between adjoining roof structures. The ties should be fixed with 3.35 mm diameter, 50 mm long galvanised round wire nails.

In extremely exposed situations, care must be taken to check that the parameters laid out in the British Standard are not exceeded, if so then a separate strapping design will be required. The standard spacing for straps is 2 m on both rafter and ceiling tie, but this may have to be reduced to provide more straps if additional restraint is required. The straps may also need to be longer, being attached to three adjacent trussed rafters rather than the standard two.

WATER TANK PLATFORMS

Another load frequently supported by the roof structure is the cold water storage tank and also possibly the heating system expansion tank. These loads must be taken into consideration when designing and constructing the roof. A check should be made with the local water authority for their precise requirements on the tank size required. The capacity of course dictates the weight of water to be supported, whilst the size may control the location of internal roof members. The capacity will usually be 230 or 330 litres. The NHBC have specific requirements for the access to and around water tanks in roof spaces and these are clearly stated in the *Registered House Builder's Handbook* part 2, clause S15 d and e. Briefly, access boarding must be provided from the loft access to the tank stand, and at least one square metre of boarding must be provided around each tank.

In traditionally constructed roofs tanks have usually been, in the past, supported directly from walls below independently of the roof, or on a stand bolted to the gable brickwork. If supported directly from the walls below a detail similar to that shown in fig. 7.19 could be used, but beams B must be designed to carry the load from the supporting walls which would themselves be replaced by beams A. The gable wall supported stand must of course have a specifically designed, probably steel bracket system which itself must be very securely fixed to the wall structure.

If the roof is to take the weight of the tanks and their supporting platform, the load will be placed invariably on the ceiling joists which are, of course, supported by binders and hangers from the purlins. All of these members will have to carry the additional weight and, moreover, careful attention must be paid to the fixings between them. It will be found necessary to use bolted connections rather than simple nails to transmit the loads from one member to another. Again a tank platform similar to that indicated in fig. 7.19 could be used with beam A being supported by hangers from the purlins above.

The bolt and connector truss roof

For the bolt and connector truss roof a separate design should be prepared for the

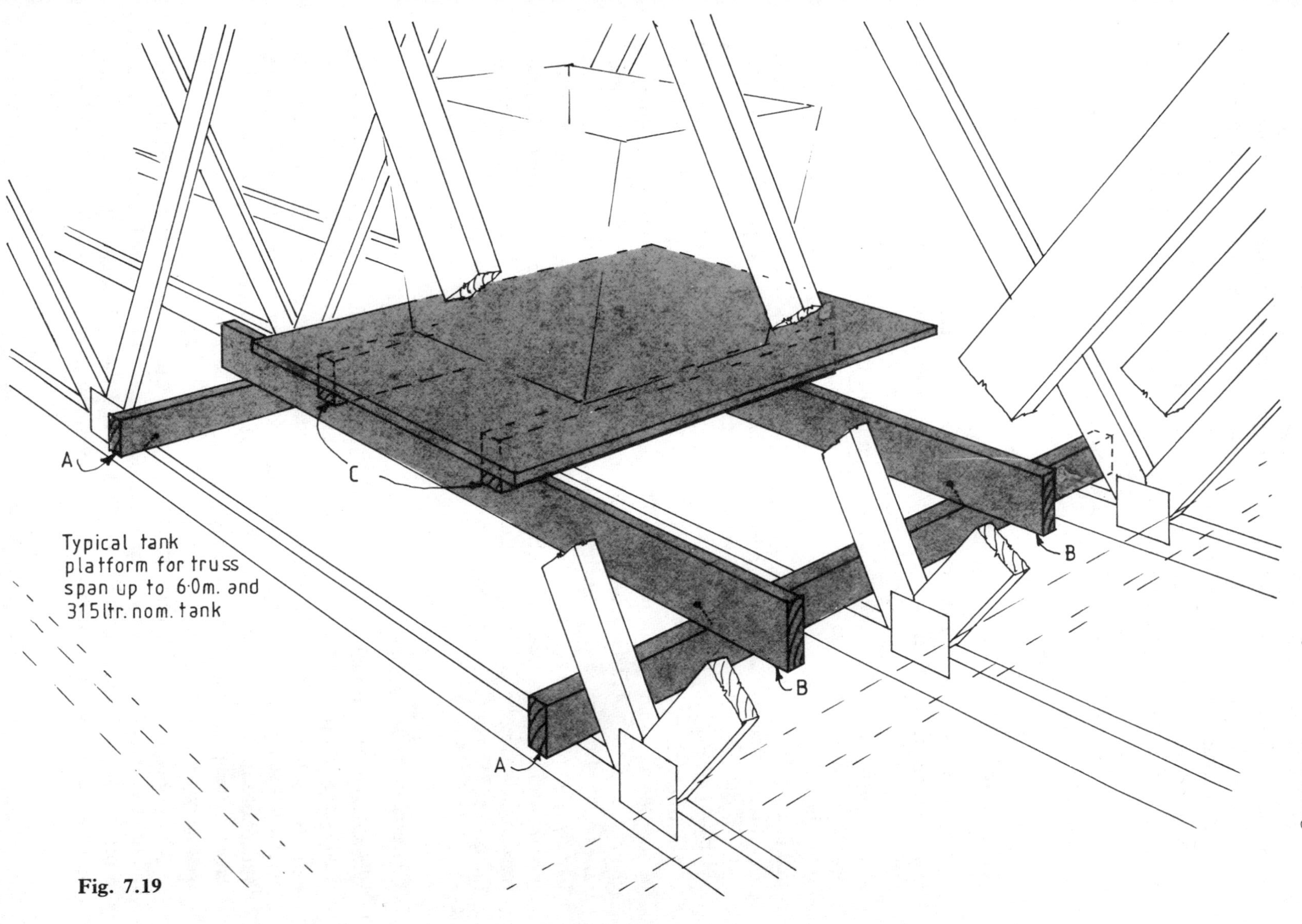
A
C
B
B
A
Typical tank platform for truss span up to 6·0m. and 315ltr. nom. tank

Fig. 7.19

tank platforms with as much of the loads as practical being carried by the principal truss itself. The TRADA designs do not allow for tank loads and therefore any tanks carried in roofs constructed using the standard design sheets must be supported from walls below. One way of overcoming this problem is to close the centres of the principal trusses in the area of the tank thus effectively strengthening the roof in that area, enabling the structure to carry the tanks independently of walls below. Design advice must be obtained from TRADA. In such instances the design shown in fig. 7.19 can again be used.

The trussed rafter roof

In trussed rafter roofs, British Standard 5268 part 3 demands that a standard tank load shall be allowed for in the design, unless the specification specifically states that no tanks are required to be carried by the roof structure. Fig. 7.19 shows a typical tank stand constructed to the guide-lines laid down by British Standard 5268 part 3, but for precise details of timber sizes reference should be made direct to the British Standard or to the ITPA *Technical Bulletin* number four, or to any of the truss rafter plate manufacturers' technical brochures. Depending on the tank size, it must be supported over three or four trussed rafters, always with bearers A as close as possible to the node point of the truss. A method of lowering the tank stand were restricted headroom is a problem is shown in the ITPA *Technical Bulletin* number four.

The question of restricted headroom must be considered when deciding upon the location of the tanks within the roof. The tank will normally be supported about the centre line of the roof truss, this generally being the largest clear void within a trussed rafter roof. Low pitched roofs bring their own problems of restricted headroom, likewise the numerous timber members in a hip end roof generally prevent tanks being placed within the hip area. Not only must access to the tanks be considered, but also reasonable space be left around the tank for initial installation and maintenance thereafter. Do not forget the additional thickness of tank insulation.

General considerations

Where showers are installed in rooms below, it is often required to raise the level of the tank as high as practical to afford head of water for the shower. The tank platform indicated in fig. 7.19 can still be used, but modified by building a rigid ply clad box-like structure around and above supports C. On the top of this box the normal water tank platform itself may be securely fixed. *Do not nail supports B to the sides of the trussed rafter webs*, they are not designed to carry such loads and the results could eventually be disastrous.

In refurbishment work, where the plumbing system is often renewed, the designer and/or installer must make sure that adequate ventilation is provided for the roof space to prevent condensation occuring on pipes to, and the tank itself. This condensation can be a cause of seriously wetting the tank platform structure, causing degradation of both platform and supporting bearers. If chipboard is used for the platform then it should be of the moisture resisting grade.

VENTILATION OF ROOF VOIDS

The ever increasing need to conserve the energy used for home heating has led to higher insulation standards in recent years, and further improvements in these standards can be anticipated in years to come. The insulation is usually placed between ceiling joists, resulting in what is known as a cold roof space, as against a warm roof space, which would be the case with insulation placed within the rafters. The following discussion assumes a cold roof space situation.

Moisture vapour rising from the home below passes through the ceiling and insulation into the cold atmosphere of the roof and will, under certain conditions, condense into water on the coldest elements within the roof space. These cold elements are invariably the metal fixings used to construct the roof and of course, in the case of a trussed rafter roof, the nail plates used for the trussed rafter assembly itself. Clearly this wetting can lead to deterioration of the metal fixings and, if continued for long periods, to the deterioration of the timbers if they are not fully preservative treated.

The solution

Two possible solutions exist, at least in theory. Firstly, prevent the moisture-laden air entering the roof space or, secondly, ventilate the moisture-laden air before it can condense and cause any harm. Examining the first option, this at first seems the most simple method, and in theory this can be achieved by placing a vapour barrier of polythene or similar vapour proof sheeting immediately above the ceiling finish, yet beneath the insulation. This would contain the moisture-laden air within the building below. In practice, however, there are numerous small holes through this vapour barrier in the form of electrical services, hot and cold water services and soil and vent pipes. There is also the problem of effectively sealing joints between the sheets of polythene used. The major drawback to this option in the writer's opinion is that a vapour barrier immediately above the ceiling finish effectively traps some of the vapour, and can lead to high moisture contents within the corners of bedrooms in particular, leading to unsightly mould growth which can be extremely difficult to eradicate once it has appeared. It is virtually impossible to contain moisture within rooms such as bathrooms, shower rooms and kitchens and therefore this first option of a vapour barrier is not a practical one.

We are therefore left with the option of ventilating the roof void. British Standard 5268 part 3 sets out minimum ventilation standards for the construction of trussed rafter roofs, but it must be remembered that these standards could well be applied to other forms of modern roof construction. The Building Regulations themselves in approved document F clause F2 'Condensation' also require the designer to take account of the possibility of condensation within the roof space. ITPA *Technical*

Bulletin number five gives details of how this ventilation should be provided in conjunction with trussed rafter roofs.

Satisfactory ventilation will be provided by designing a minimum gap of 25 mm along at least two opposite sides of a roof where the pitch does not exceed 15°, or 10 mm for roof pitches above 15°. Furthermore, when a monopitch roof is being considered, a duopitch roof in excess of 20°, or 10 m span, consideration should be given to providing further continuous ventilation at the ridge. There are many ways of effectively providing this ventilation, mostly via the soffit in the form of slots or holes covered with an insect-proof gauze. Having introduced the airflow into the soffit void, care must be taken to prevent insulation blocking the space between the rafters, thus preventing this flow of air into and through the roof space. In addition, therefore, some method of controlling the insulation must be introduced.

Proprietary systems

The two leading roof tile manufacturers in this country, namely Marley and Redland, both have their own systems for providing ventilation at eaves level and at ridge roof spaces. Both systems use lightweight plastic type mouldings to provide the ventilation, insulation control and the necessary insect screening. Both companies produce systems where the ventilation is provided on top of the fascia and not through the soffit by, in Redland's case, a single plastic moulding, and in Marley's case a set of mouldings which at one and the same time provide ventilation, support the roofing underlay felt to prevent ponding and control the insulation. Both systems have a ridge ventilation method which is 'dry fixed' meaning that it does not depend on traditional cement mortar for secure bedding.

Other companies manufacture individual ventilators and insulation controllers, the latter usually designed for use with trussed rafters at the standard 600 mm centres. All provide components complying to the necessary requirements, but careful specification of the individual item is necessary to ensure that the correct air gap for the pitch being used is achieved. Fig. 7.20 shows the essential features of eaves ventilation and insulation control.

The attic roof

The attic roof of course has three roof voids, those at either side at low level and the triangular roof void above the attic room. Airflow must again be introduced at low level, i.e. at the eaves, and allowed to flow unobstructed from the lower roof void up between the rafters to the upper roof void, where exhausting from the ridge is particularly important because of the usually steep pitch nature of the attic roof.

The area of rafter which forms the sloping ceiling of the room must be carefully considered. When attic trussed rafters are used then this rafter may well be 200 mm deep which, when 100 mm of insulation is used between the rafters, should allow adequate ventilation space above the insulation from one roof void to the next. However, unless the designer can be absolutely certain that the insulation will not slide

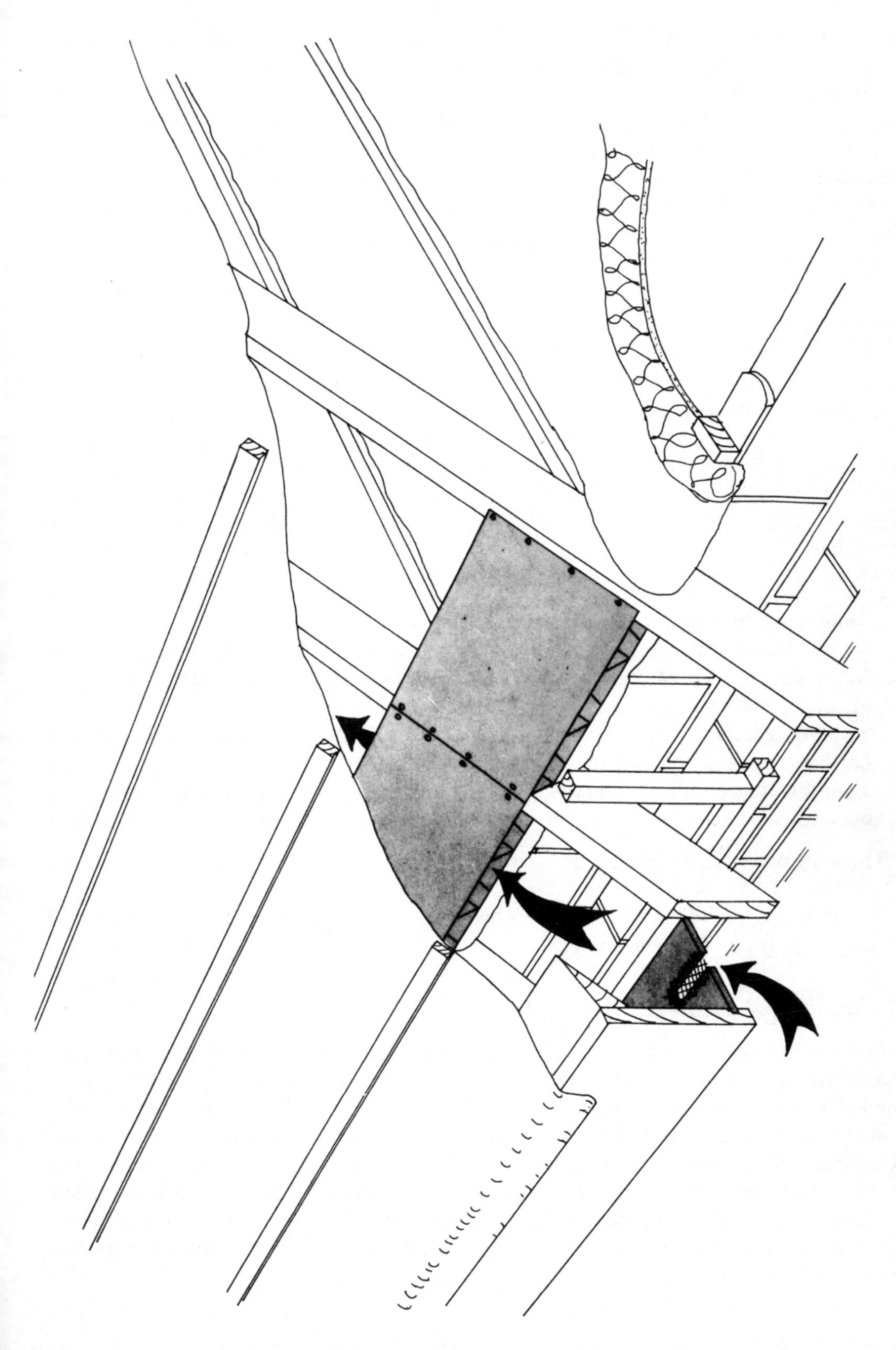

Fig. 7.20

down the sloping ceiling, some controls may need to be introduced to ensure that adequate ventilation space of 25 mm between each rafter is maintained.

On traditionally constructed attic roofs where the timbers may be smaller, or in the type of construction illustrated in figs 5.19 and 6.13, the timber used for the rafters will be relatively small. To ensure that adequate insulation thickness and air space are maintained, it may be necessary to fix a minimum rafter depth of at least 150 mm thus giving room for 100 mm insulation, an insulation controller and a 25 mm air space. In the next section, this subject of insulation control is again highlighted in the discussion on sheet material used for stability bracing in attic roof construction.

BRACING

Any roof constructed of simple timber members or manufactured components is about as stable as a row of dominoes standing on their ends. Similarly, any pressure on the end of these components will result in a 'domino effect', each toppling on to the other until possibly total collapse occurs. The roof clearly has to be braced against this effect and it is not adequate to consider the brick and block gable ends satisfactory for this purpose. The section on gable end restraint should have emphasised this point adequately.

There are three main types of bracing to be considered:

(1) Temporary bracing required to stabilise the roof whilst it is actually being worked on and in the construction stage, leaving it secure during breaks in construction. Many partly built roofs have been damaged by sudden storms when left in an unstable condition between work periods
(2) Stability bracing, required to brace the individual timbers in position in the finished roof and to keep the roof itself upright
(3) Wind bracing, required to enable the roof structure to resist loads imposed upon it by wind acting on gable ends, or in some instances on the supporting walls below

These three types of bracing apply to all forms of roof construction (refer to fig. 7.17).

A fourth type of brace, the 'Web Compression Brace', may be required by the truss designer. See page 76 for additional detail.

The traditional cut roof

Refer now to illustration fig. 7.21. On traditionally constructed roofs little temporary bracing is required if the purlins are well fixed at their support points. This means of course ensuring that supporting brickwork has not only been constructed but that the cement mortar is adequately cured, otherwise damage to the wall could occur from the movements of constructing the roof above it. Although the illustration shows the trussed rafter roof detailed in chapter five, the essentials of bracing of course remain the same in any roof form. No specific guidance exists for the bracing of traditional roofs, but the recommendations laid down in British Standard 5268 part 3 for trussed rafter roofs could well be followed with few modifications.

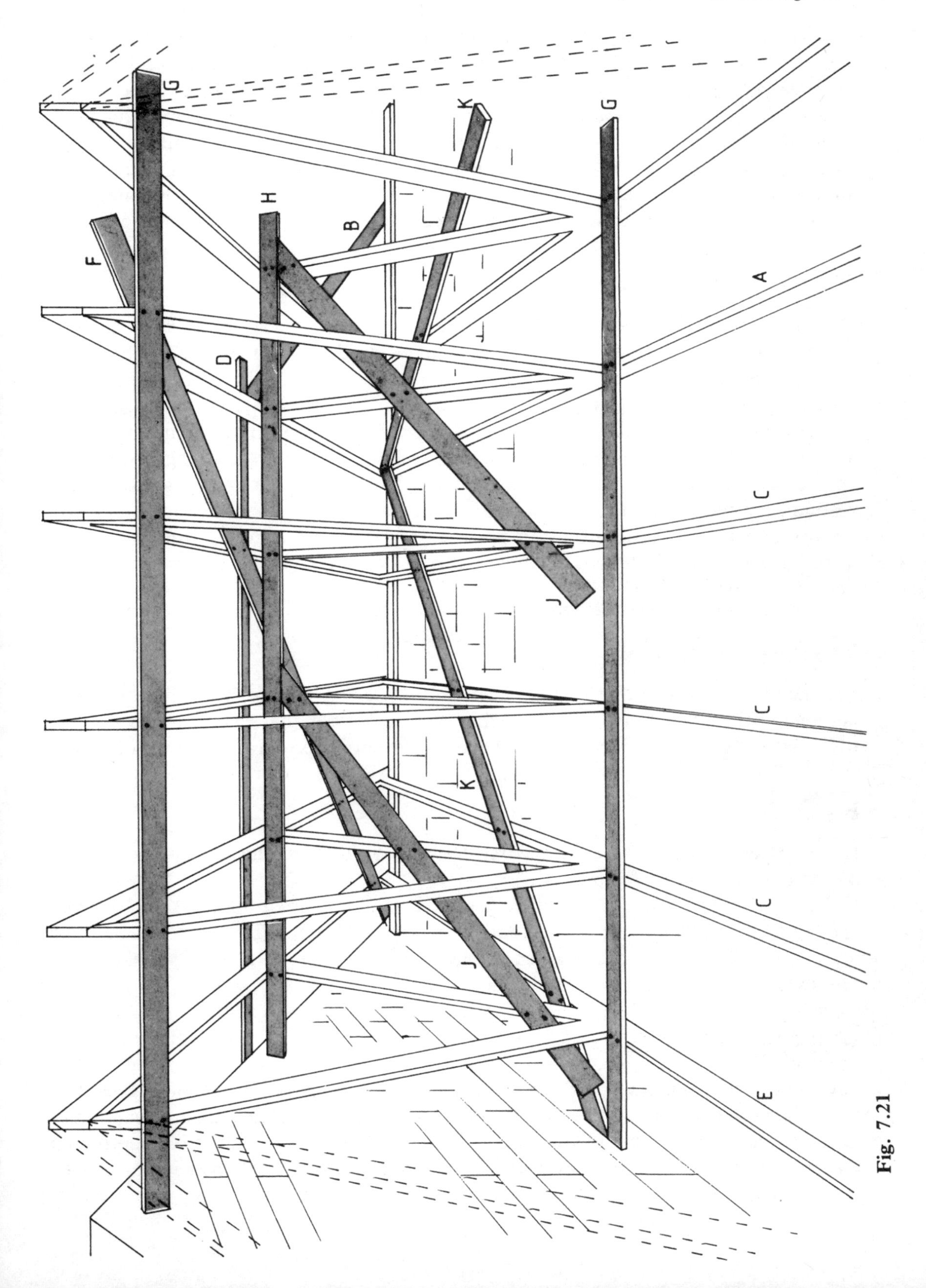

Fig. 7.21

It is advisable to fit the diagonals B, the binders G being replaced by the ridge board and the ceiling joist binders themselves. Stability bracing can be achieved by adding diagonal brace F with brace J being fitted from the binder diagonally across the hangers to the purlin. Because the hangers may well be spaced further apart than with a trussed rafter system this member should be more substantial, probably 50 mm × 100 mm in section. If the purlin strut is used as illustrated in fig. 3.5 this additional brace will not be required. Brace H in fig. 7.21 is of course the purlin itself in a traditional roof. Finally the ceiling joist diagonal brace K should be fitted as indicated.

The bolt and connector truss

The principal trusses are significantly heavy and great care must be taken to temporarily stabilise the first pair of these trusses fixed in position. Again referring to fig. 7.21, braces B should be installed as early as possible but should be of 50 mm × 100 mm cross-section and well nailed to both the trusses and the wall plate. Furthermore, a prop should be fitted under the ridge collar down to the first floor joist system (or floor if a single-storey building) below. Because of its length this prop should be a scaffold tube fitted with a suitable clip to lock it on to the ridge collar, and restrained at the bottom by timbers laid flat and nailed to the floor joists, or fitted tight against a cross wall. Apart from this precautionary measure resulting from the weight of the trusses, the other items of bracing can be considered as for the traditional roof described above.

The trussed rafter roof

Bracing for the trussed rafter roof has been fully described in Chapter 5 because of its inclusion in British Standard 5268 part 3 which was analysed in that section, and because of the trussed rafter's dependence upon the bracing system for its overall structural integrity.

Hip roofs

Hip roofs are essentially well braced structures by virtue of the hip board diagonal braces inherent in the design.

Again referring to fig. 7.21, binders G and H should be fitted to the section between the hip ends and as far into the hip ends as possible. Diagonal K should also be fitted and, unless the distance between the hip peaks is very long, diagonal F should not be required.

Valley intersecting roofs

Each roof section should be treated individually with the guidance set out above. The valley boards themselves add extra diagonal stiffening in that area. The discontinuity of tile battening under the abutting roof should be replaced by timber binders placed on top of the common rafters as described in trussed rafter valley roofs in Chapter 6.

Attic roofs

Attic roofs, because of their great height and the inherent large void forming the rooms, present their own particular problems with regards to bracing. The trussed rafter attic has been dealt with in some detail in Chapter 5. It is unlikely that an attic will be constructed using bolt and connector principal trusses, the notes below therefore apply to traditionally constructed attics and also generally to trussed rafter attic roofs, although in the latter case specific designs for bracing should be obtained from the truss designer.

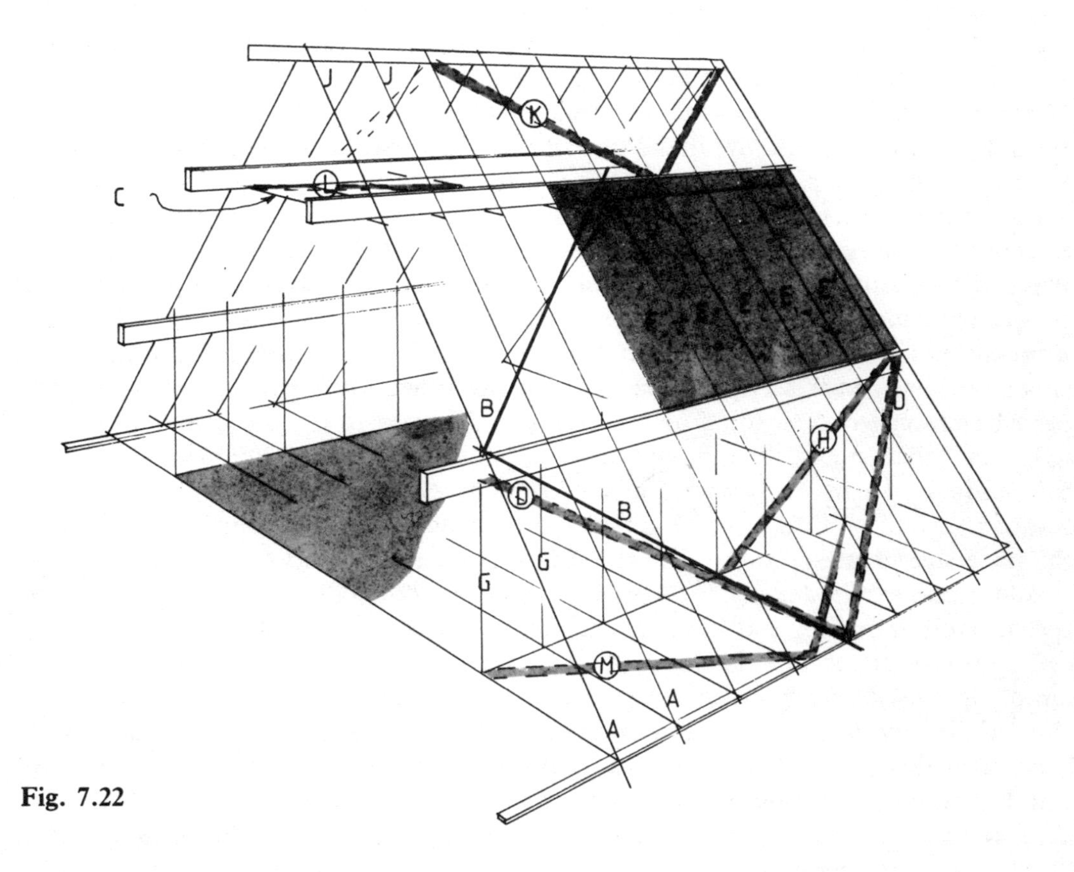

Fig. 7.22

Fig. 7.22 should be referred to. For stability reasons, the gable walls will probably not be constructed above the upper purlin level at the time roof construction starts, and it is assumed in this description that the long rafters will not be available in one length, and have been split into the upper and lower rafter sections, namely A and J respectively. The description also assumes that the purlins and floor joists are fixed in position.

Construction should commence by fixing six of the lower rafters A from wall plate to upper purlin on both sides of the roof. Temporary diagonals B should now be fixed in place on both sides of the roof, securely fixed to each rafter passed. Collars or ceiling ties C should now be fitted to stiffen the upper purlins. Next complete the commons to the lower part of the roof and the collars. Diagonal braces D should next be fixed to the underside of rafters A. Next fix uprights G between purlin and floor joist to form the walls of the room. Care should be taken to set these uprights or studs at centres appropriate to the wall covering board to be used. Diagonal H should now be fixed across the end five or six uprights, nailing this to the void side of the stud, and nailed to each stud it passes.

Having stabilised each lower section of the roof, attention must now be turned to the space between the purlins, to ensure a solid structure before completing the roof to the ridge. To avoid a diagonal between the purlins on the underside of rafters A a sheet material E, which is suggested to be 9 mm or 12 mm regular sheathing ply, should be tightly fitted between the common rafters A for approximately five rafter spaces away from the gable end. This ply should be nailed to battens fixed to the side of rafters A at their top edge. The plywood, being fixed to the underside of these battens, will then act as insulation controller in this sloping part of the ceiling to the roof below, yet maintain an air space above it and below the felt underlay to the tiles. The batten depth therefore must be a minimum of 25 mm. This ply diaphragm, provided it is adequately nailed, will impart great stiffness to this otherwise unbraced part of the roof, the ply panels being fitted on both sides of the slope and at both gable end situations. The upper section of these panels may have to be left off until the upper rafters J are nailed in position on the purlin and lapped to the top of rafters A.

Finally rafters J and their ridge can be fitted, the rafters being birdsmouthed over the purlin in the usual manner and well lapped and nailed to the top of rafters A. Racking sheathing E can now be completed. Diagonals K and L can now be fitted to the underside of the rafters and to the top of the ceiling collars C.

All timbers mentioned above for bracing can be of 22 mm × 97 mm section, with the specification for nailing and lapping, if necessary, as set out in Chapter 5.

The final brace for an attic roof is of course the floor boarding or sheeting applied on top of the floor joists between the uprights G. This floor diaphragm is subjected to more loading by the structure above it than with a conventional two-storey building. It is essential therefore that the floor joists are solidly bridged between them, or fitted with herringbone strutting, and that supports are provided to carry the floor boarding, particularly at the junction with the wall G. Correct nailing of any boarding used for the floor must be used; if chipboard is used this would mean annular ring shanked nails and gluing of the tongued and grooved joint between the boards.

The effect of openings on bracing

All attic roofs will have some openings in the sloping part of the ceiling, either in the form of roof lights or dormers. The effect of these openings on the bracing will naturally depend upon the frequency and size of opening in either or both roof's slopes in relation to the overall length of the building. Clearly the provision of adequate bracing in the form of the panels E must be considered and if this cannot be provided immediately adjacent to a gable wall, then continuity between the panels near the wall and those repositioned because of an opening must be maintained to allow the two separate areas of bracing to act as one.

Stairwell openings within the main body of the building are unlikely to cause any signifiant decrease in the effectiveness of the floor diaphragm. Those staircases constructed parallel with the floor joists and immediately adjacent to a gable wall, however, may present stability problems for that gable wall which will invariably be supporting one of the purlins carrying the roof. A means of providing an alternative to brace H must be found and additional attention may be needed to gable end restraint in such instances, above and beyond that dealt with in the section on gable wall restraint in this chapter.

EAVES DETAILS

The eaves of a building vary greatly in design throughout the UK, and to some extent are an architectural detail rather than a structural requirement. The function of the eaves is of course to close off the ends of the rafters and, where a generous overhang is provided, to protect the building to a certain degree below. The traditional large overhang associated with most thatched roofs proved excellent protection to the heads of doors and windows below.

The functions to be considered in the design of eaves are therefore as follows:

(1) To effectively close off the space between the rafter feet
(2) To provide a means of ventilation for the roof
(3) To provide protection for the building below if required
(4) To provide support for the rainwater drainage system
(5) To provide support for the tile underfelt
(6) To provide support for the soffit if required

One of the most important features mentioned above is the support of the tile underfelt. Fig. 7.23a shows the problems of underfelting unsupported, being allowed to sag without support between the rafters, thus allowing ponding, with the resulting degradation of both fascia and soffit and possibly the top of the wall structure itself. Adequate support must be given at the bottom of the roof slope for the felt to avoid this ponding, this being achieved in the form of a thin sheet material applied to the top of the rafter feet or sprockets if provided, or in the form of a continuous triangular fillet fixed to the top of the rafter feet. This detail (fig. 7.23b) allows any water which may have penetrated through the tile to run down the roof slope into the gutter in the normal manner.

The next important aspect is to detail the eaves allowing adequate ventilation, and simple methods to achieve this are indicated in figs 7.24a-e.

Fig. 7.24a shows a detail with no overhang, care being taken not to fix the fascia tight to the wall, although with a ventilation system shown in fig. 7.24d the fascia could be fitted directly to the wall if necessary. Fig. 7.24b shows a typical overhang

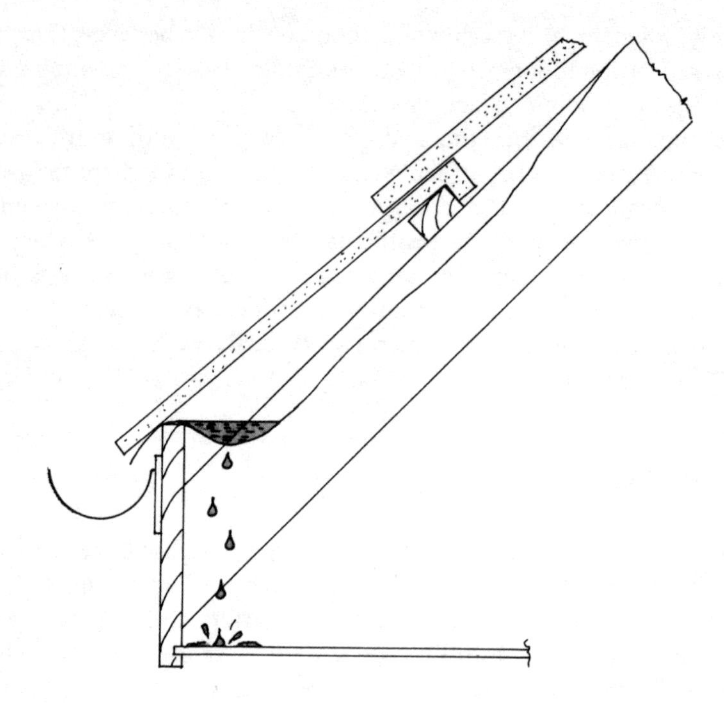

Fig. 7.23a

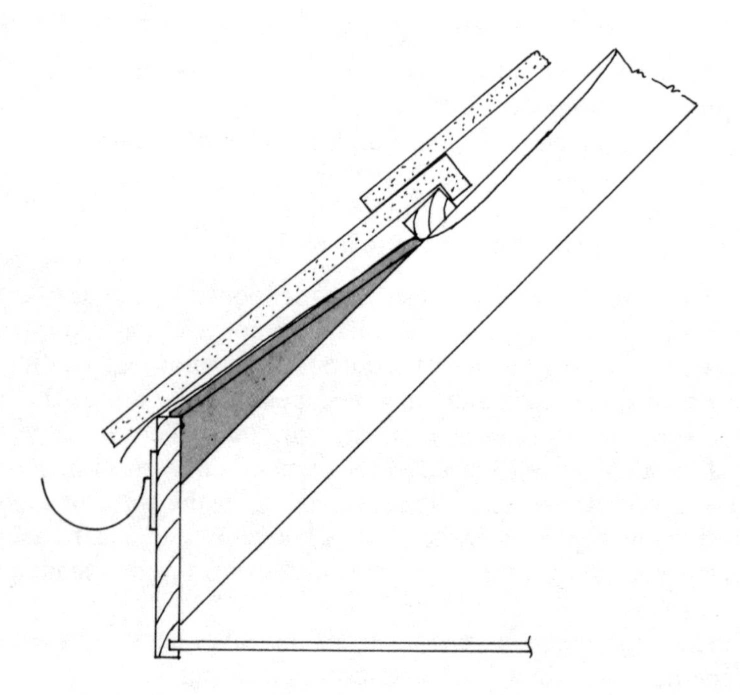

Fig. 7.23b

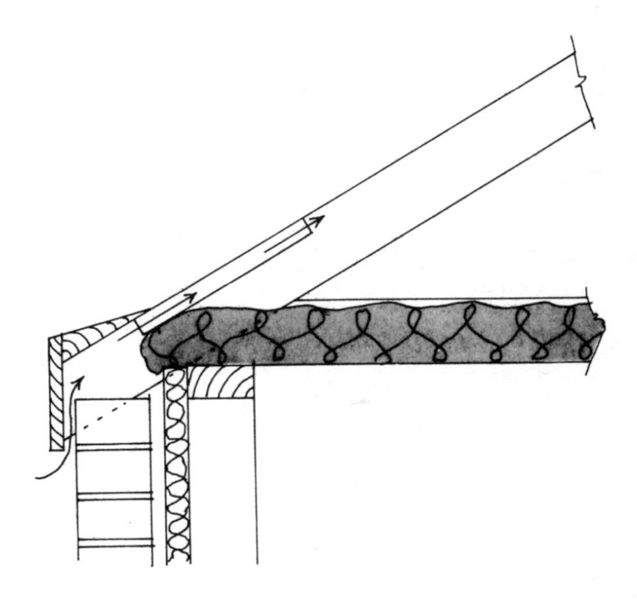

Fig. 7.24a

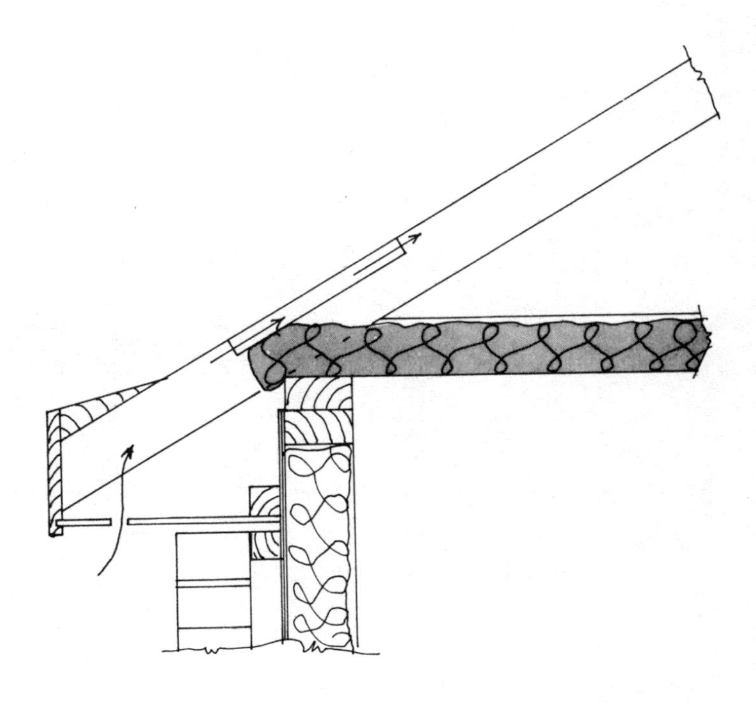

Fig. 7.24b

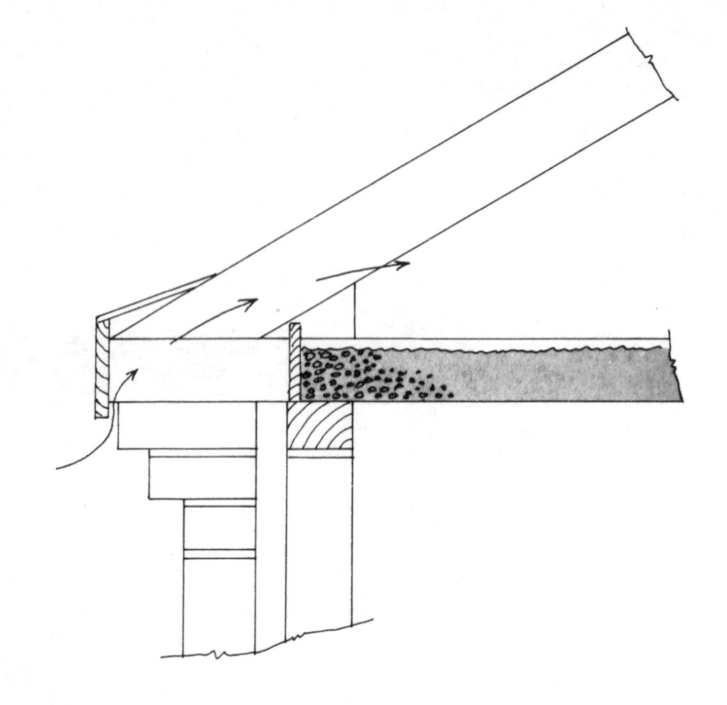

Fig. 7.24c

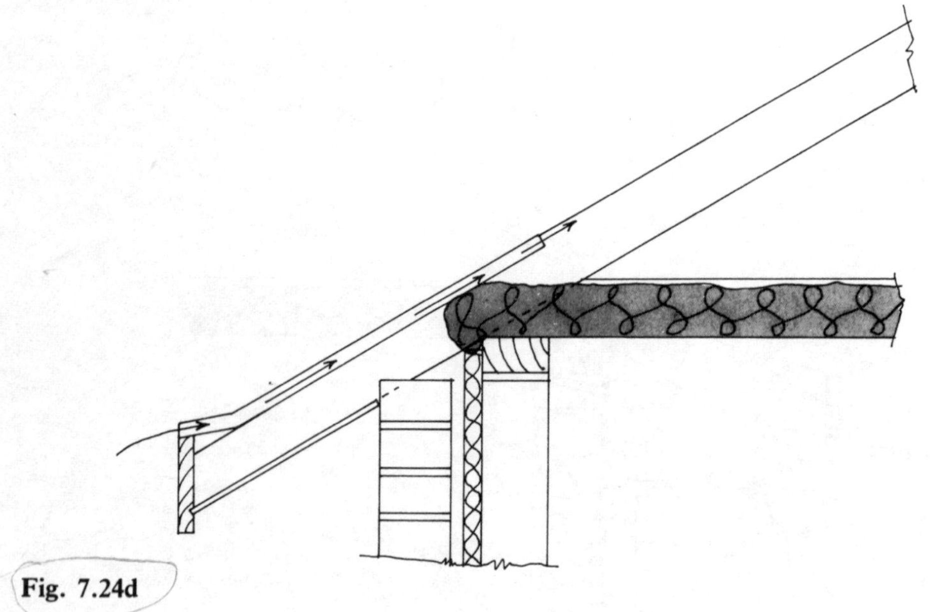

Fig. 7.24d

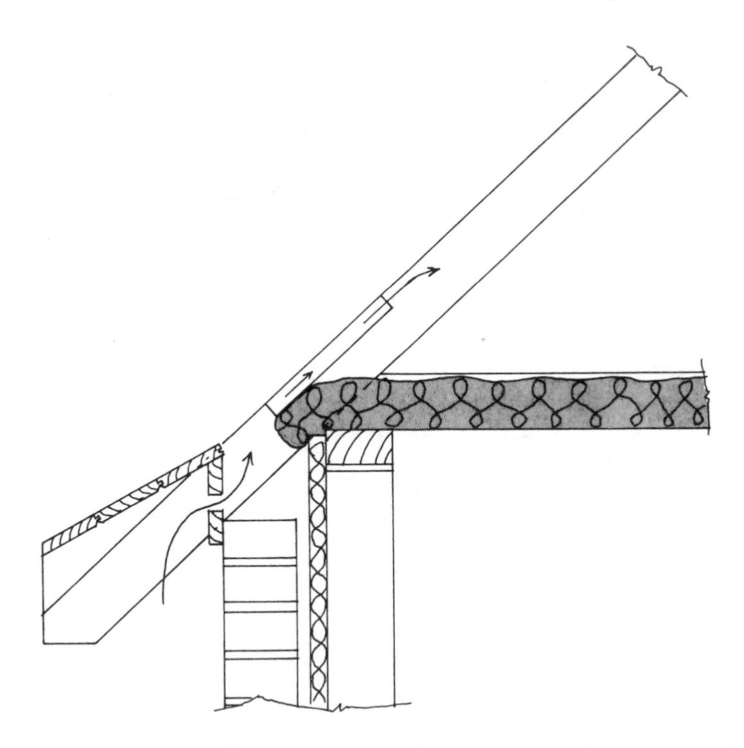

Fig. 7.24e

with fascia and soffit, this particular detail indicating a timber framed structure, care having been taken in this instance to show a gap between the soffit and the top of the brickwork to allow for the differential movement between it and the timber structure. Fig. 7.24c shows a corbel eaves detail with no soffit and with the junction between rafter and ceiling tie taken beyond the outside of the wall. This particular detail would impose certain structural problems for all types of roof construction, and may require a blocked heel or additional top chord, should trussed rafters be specified. This particular detail indicates a loose fill insulation with a timber board controller to prevent the insulation spilling through into the cavity, or across the cavity blocking the ventilation space.

Fig. 7.24d shows one of the proprietary combined ventilators and insulation controllers fitted on top of the fascia. This particular detail also indicates a sloping soffit fitted directly to the underside of the rafters. Fig. 7.24E indicates exposed rafter feet with ventilation provided by slots between the infills between rafters.

The above illustrations show only a few of the many variations on design imparting individuality to any building. The only two details likely to give any structural problems are those indicated in 7.24c because of its cantilevering effect for the truss, and any of the details where the rafter overhang is excessively long. In general this would mean beyond 700 or 800 mm, depending on the rafter depth. The use of the triangular sprocket piece on top of the rafter foot will not aid its strength in this respect, unless of course it is carried up the rafter well beyond the wall plate position.

Fascias, barge boards and soffits should always be preservative treated in accordance with building regulations and NHBC requirements and should be given one coat of either paint or stain prior to fixing.

The soffit boards need not be preservative treated for they are generally not exposed to the weather, although in the writer's opinion it is desirable to do so if softwood tongued and grooved boarding is used in an exposed eaves detail, such as fig. 7.24e. It is normal to support the soffit at the fascia by fitting it into a groove in the back of the fascia, and on light timber softwood framings on the wall side of the building. Fig. 7.20 illustrates a well framed soffit support system.

Exposed rafter feet

Exposed rafter feet tend to be a fashion feature, but are very common in rural areas where new buildings need to blend architecturally with those older dwellings surrounding them.

Where trussed rafters are used in conjunction with a detail such as that indicated in fig. 7.24e, it must be borne in mind that the timber will be planed not sawn, as will have been traditionally the case, that it is likely to be stamped with bright red or other colour stress grade marks, and that for economy reasons the trusses (and therefore their exposed rafter feet) are likely to be spaced at 600 mm centres compared to the more normal 400 mm centres of the traditional buildings.

Assuming that the centres of the trussed rafters are acceptable, the grade marks may be overcome by the use of a dark stain or indeed any paint system, but light stains will not be adequate to conceal the marks. To overcome the problem of centres and the fact that the rafter feet are planed, the detail indicated in fig. 7.25 could be used to give a more authentic eaves detail. This allows the economy of the trussed rafter to be placed at 600 mm centres. It also allows the architect to choose the precise centres at which he would like to place the rafter feet and allows him to use sawn timbers of possibly a heavier section than the timbers used for the trussed rafter construction. This particular detail also solves the problem of irregular trussed rafter spacings which invariably occur around hip end roofs. It is suggested that the dimension X should be equal on either side of the wall plate position. By birdsmouthing the supplementary rafter feet over the wall plate, a deeper rafter than that used for the truss rafter itself may be used if desired.

TRIMMING SMALL OPENINGS

The openings dealt with here are those required for small flues and loft access hatches, as well as the smaller openings for roof windows. Large openings have been detailed in the relevant preceeding chapters.

In designing the roof, careful consideration should be given to the location of any flue or chimney passing through the roof void, and its likely effect on the structure. Similarly the loft hatch should be considered, particularly in relation to the spacing of trussed rafters. Roof windows for fenestration reasons may have to take priority over

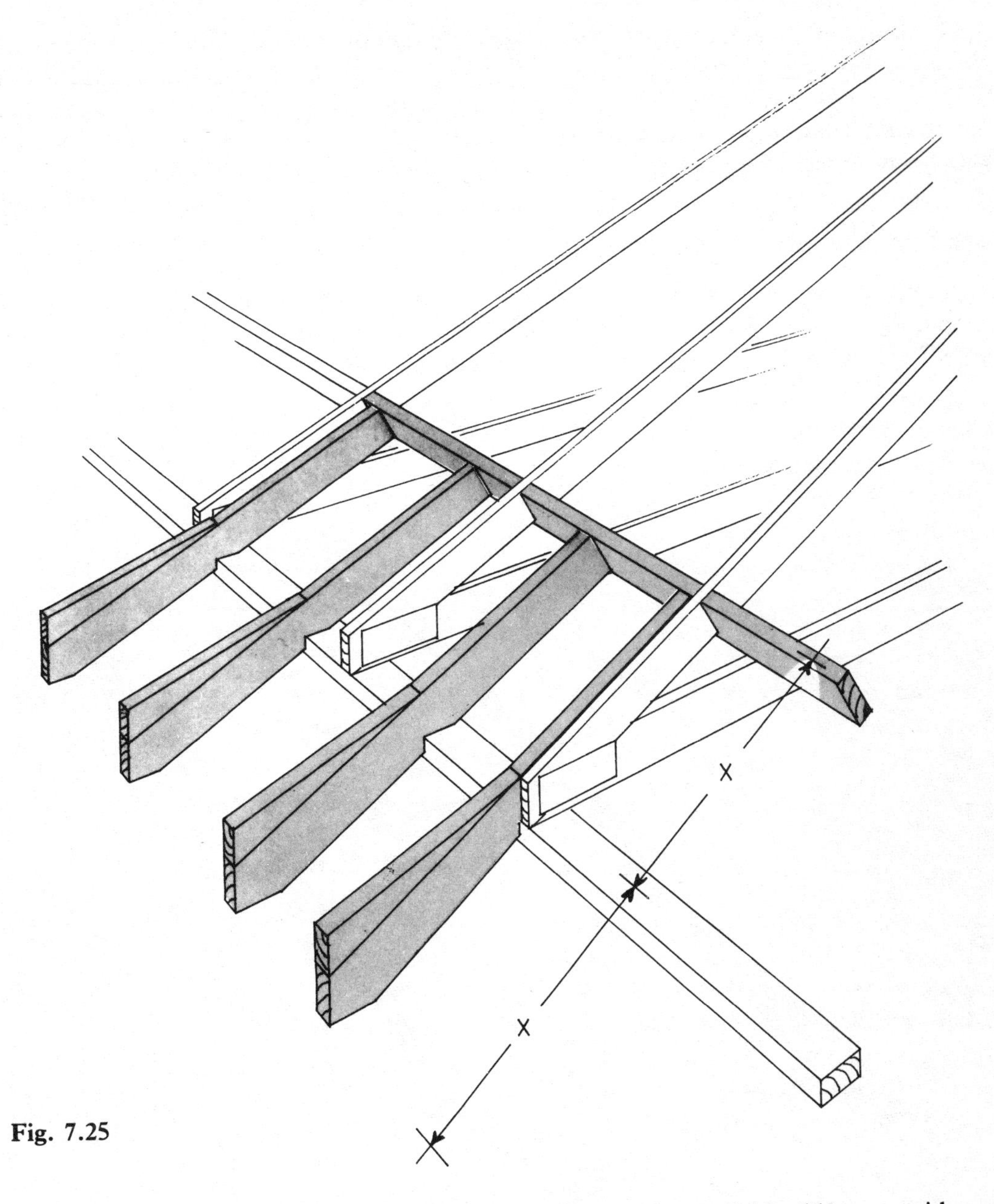

Fig. 7.25

the structure, but again with consideration, the readily available 550 mm wide roof windows should be used if possible to fit neatly between trussed rafters, thus achieving ultimate economy in the structure by avoiding trimmed openings.

Where brick flues pass through roofs, the Building Regulations stipulate that 40 mm minimum must be allowed between the structure and the brickwork face. Floor boards, skirtings, tile battens, etc., may of course touch the brickwork but not the actual structure itself. Please refer to the relevant Building Regulation clause for precise details with regard to different classes of appliance and flue sizes.

The Building Regulations do not stipulate minimum size for loft hatches, but the NHBC do in their *Registered House Builder's Handbook* part two, clause S15B. This requires a minimum size of 550 mm clear in any direction. The clause also states that the opening should not be directly over stairs or in any other hazardous location, such as close to eaves where headroom in the roof space is limited.

Traditional roofs

In traditional or bolt and connector roofs, small openings may be created in either the rafters or ceiling planes by simply doubling the rafters on either side and trimming, as indicated in fig. 7.26. This would be considered suitable for openings up to 1.2 m

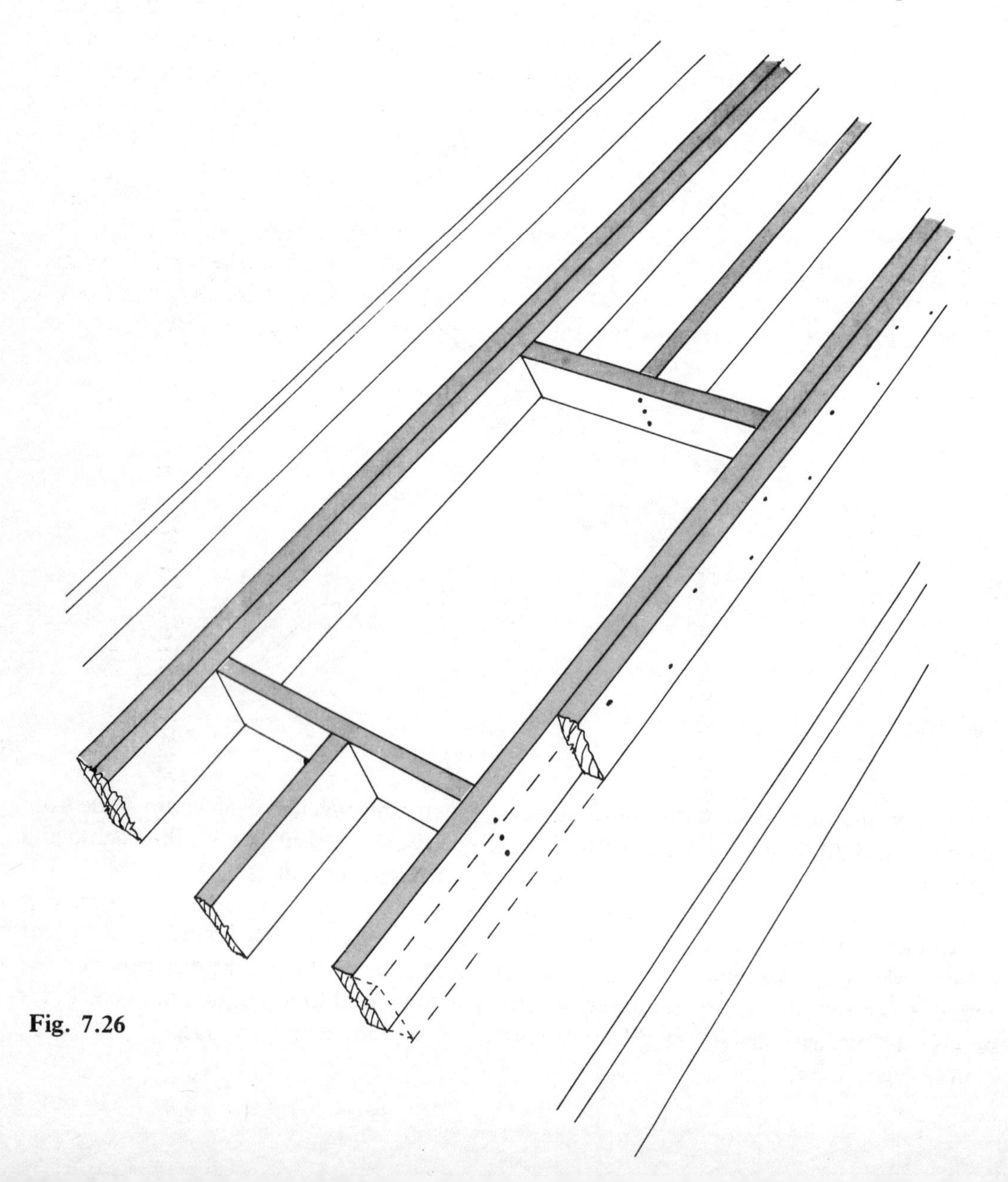

Fig. 7.26

wide, and for a length the maximum distance between two purlins or purlins and plate. Larger opening must be dealt with as set out in chapter three.

Trussed rafter roofs

Trussed rafters of course are designed to work at 600 mm centres (or some other specified dimension), and these spacings should not be increased without adjusting either the design of the roof truss itself, or the spacing on either side of the opening created. On no account should a trussed rafter be cut. Only two truss plate manufacturers appear to allow this cutting on site, these being Messrs Bevplate and Twinaplate, who give specific trimming details in their design manuals.

British Standard 5268 part 3 on pages 24 and 25 gives details of openings for chimneys and hatches. Clause 30.4 sets out details for the maximum spacing between trimming trussed rafters. Using the British Standard's lettering, the standard truss spacing equals a, the spacing between trimming trusses and the adjacent standard truss equals b, with the distance between the centres of the trimming trusses being c. This gives a formula of $c = 2a - b$. The nominal opening c is not that which the designer would need to know, so let the *actual* opening between the trimming trussed rafters be w. Assuming then a truss thickness of t, the actual opening width between trussed rafters becomes $2a - b - t$. To find the maximum opening width permissible for a truss spacing $a = 600$ mm, with a truss thickness $t = 36$ mm, we have $2 \times 600 - 36 - 36 = 1128$ mm. The dimension b must be equal to the truss thickness t, because to give the widest opening the trimming truss must be immediately adjacent to the last standard truss.

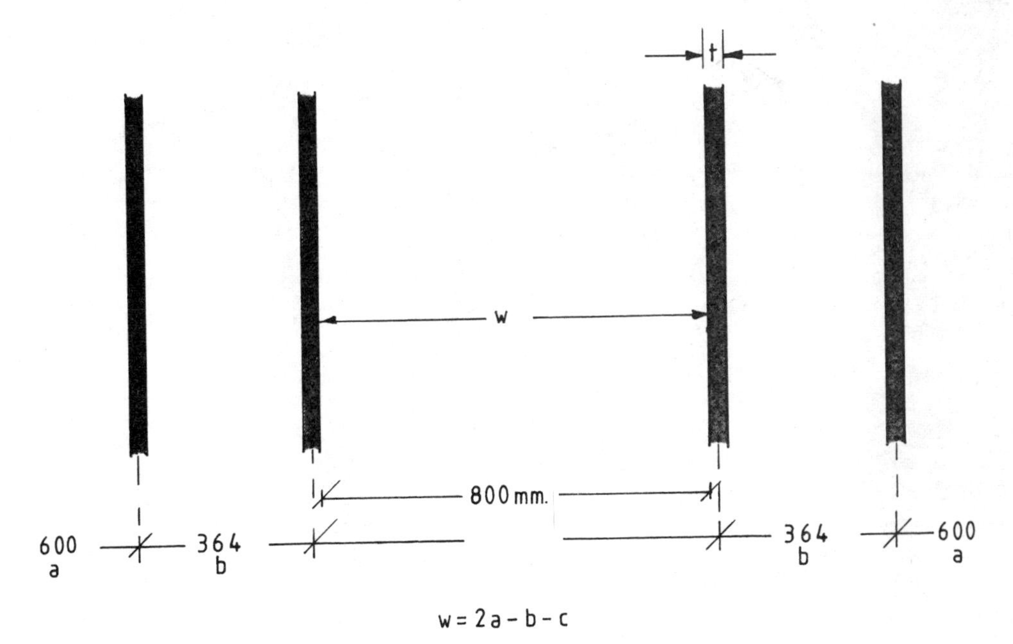

Fig. 7.27

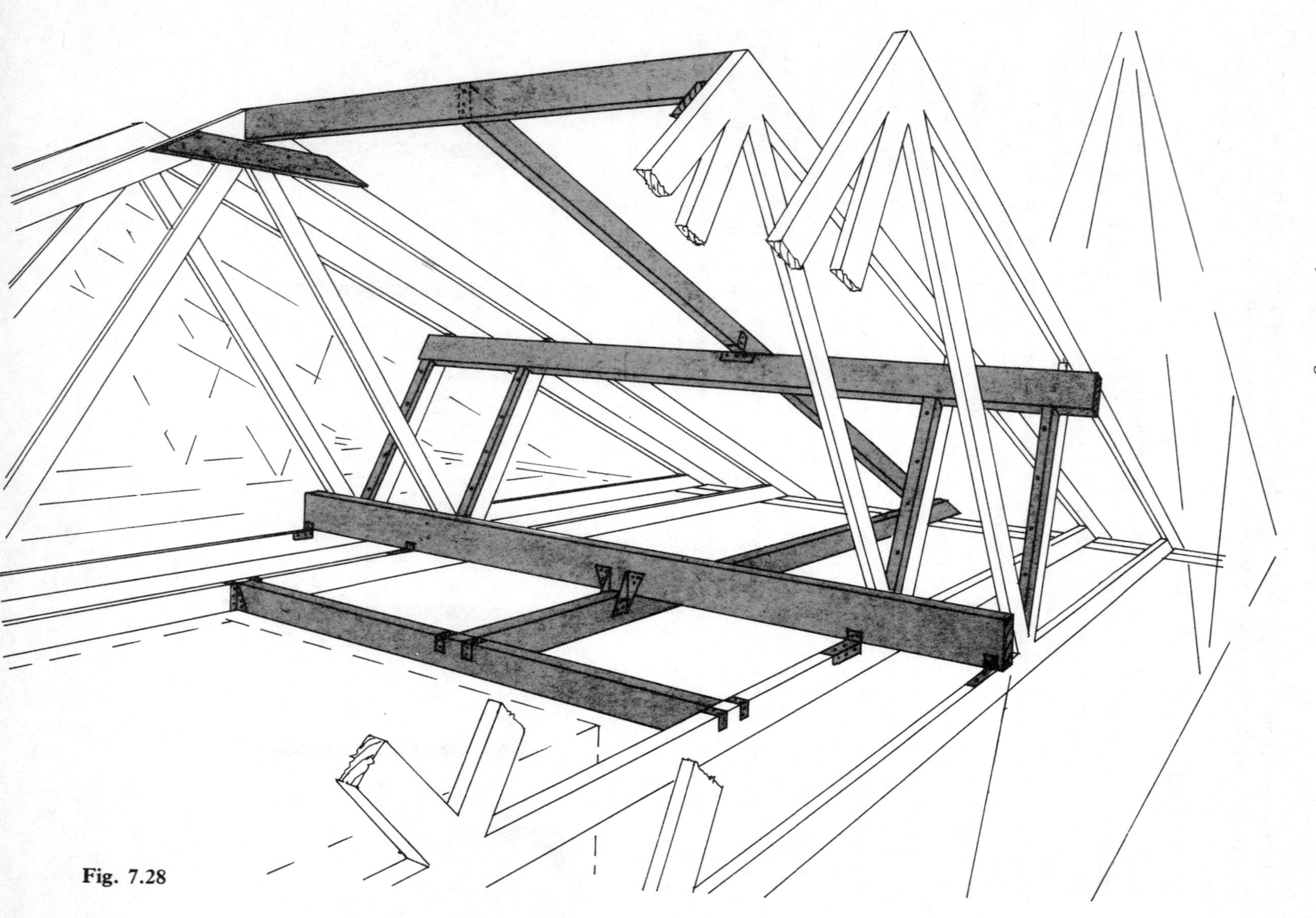

Fig. 7.28

Let us now assume that we need an actual opening of 800 mm, and w = 800, a = 600 and t = 36. Substituting in the formula above we have 800 mm = 2 × 600 − b − 36, b = 1200 − 800 − 36, b = 364 mm. The setting out of this truss would then be as shown in fig. 7.27.

INFILL

Infilling between the trussed rafters should be carried out with timbers of similar size to those for the trussed rafter members themselves supported on the wall plate, a purlin which will effectively be a heavier piece of timber nailed to the webs in the position of brace H (fig. 7.21), and to a timber member nailed between the trimming trusses on either side of the opening concerned. All of this assembly is carried out using galvanised nails, but for the 'trimmers' supporting infill between trusses it is strongly recommended that these are fixed to the trimming trusses with small metal framing anchors. Fig. 7.28 illustrates a typical chimney infill situation.

BIBLIOGRAPHY AND REFERENCES

BRITISH STANDARDS

BS 1282: 1975 *Classification of wood preservatives and their methods of application.*
BS 1579: 1960 *Connectors for timber.*
BS 4072: 1974 *Wood preservation by means of water-bourne copper/chrome/arsenic composition.*
BS 4169: 1970 *Glued-laminated timber structural members (including AMD 768 and AMD 3453).*
BS 4471: Part 1: 1978 *Dimensions for softwood. Basic sections.*
BS 4471: Part 2: 1971 *Dimensions for softwood. Small resawn sections.*
BS 4978: 1973 *Timber grades for structural use* (being rewritten).
BS 5250: 1975 *The Control of Condensation in Dwellings.*
BS 5268: Part 2: 1984 *Structural use of timber code of practice, permissable stress design, materials and workmanship.*
BS 5268: Part 3: 1985 *Code of practice for trussed rafter roofs* including amendment No 1 1988.
BS 5268: Part 4: section 4.1 1978 *Fire resistance of timber structures.*
BS 5268: Part 5: 1977 *Preservative treatments for constructional timbers.*
BS 5450: 1977 *Sizes of hardwoods and methods of measurement.*
BS 5534: Part 1: 1978 *Code of practice for slating and tiling (including AMD 2734 and AMD 3554).*
BS 5707: Parts 1, 2 and 3: *Solutions of wood preservatives in organic solvents.*

REGULATIONS AND CONTROLS

The Building Regulations 1985, statutory instrument and approved documents complete.
The National House-Building Council (NHBC) registered house builders handbook part II technical requirements for the design and construction of buildings.

USEFUL REFERENCES

TRADA (The Timber Research and Development Association). Head Office, Hughenden Valley, High Wycombe, Buckinghamshire HP14 4ND; 0240 24 3091. Regional Offices at

Scotland; Stirling Enterprise Park, John Player Building, Players Road, Stirling FK7 7RS
North West and Northern Ireland; 115, Portland Street, Manchester
North and North East; Manchester House, 48, High Street, Stokesley, Middlesborough, Cleveland TS9 5AX
Midlands; 91, High Street, Evesham, Worcs WR11 4DT
East; 149, St Neots Road, Hardwick, Cambs CB3 7QJ
Bristol Channel and South Wales; 91, High Street, Evesham, Worcs WR11 4DT
London and South East; Hughenden Valley, High Wycombe, Bucks HP14 4ND
South West; 5-7, South Street, Wellington, Somerset. TA21 8NR
Irish Republic Office; 10, Orwell Road, Rathgar, Dublin 6.

TRADA Publications

TRADA wood information sheets – the complete set, but in particular the following:

1-6 Introduction to the specification of glued laminated members.
1-10 Principles of pitched roof construction.
1-14 Building Regulations for the conservation of fuel and power.
1-17 Structural use of hardwoods.
1-20 External timber cladding.
2/3-5 Guide to the specification of structural softwood.
2/3-9 Mechanical fasteners for structural timber work.
2/3-15 Basic sizes of softwoods and hardwoods.
2/3-16 Preservative treatments of timber – a guide to specification.
4-7 A guide to stress graded softwoods.
4-12 Care of timber and wood based materials on building sites.
4-14 Moisture content in wood.
DA 1 Design examples to BS 5268 Part 2.
DA 2 Span tables for domestic purlins.
DA 3 Span tables for floor joists.
DA 4 Span tables for nailed ply-box beams.
DA 5 Span tables for glued ply-box beams.
DA 6 Joist span tables for domestic floors and roofs. Processed timber sizes.
TBL 35 Visual stress grading.
TRADA roof design sheets as listed in Chapter 4, fig. 4.3 of this book.

The Swedish Finnish Timber Council

21/25 Carolgate, Retford, Notts DN22 6BZ

All publications, but in particular the following:

The sawn timber and products.
Stress graded to BS 4978.
Floor joist and flat roof span tables for one family houses.
Load span tables for joists.
Domestic floor joists.
Roof joists.

COFI (Council of Forest Industries of British Columbia), Templar House, 81 High Holborn, LONDON WC1V 6LS

ITPA (International Truss Plate Association), PO Box 44, HALESOWEN, West Midlands.

All technical bulletins and publications relating to trussed rafters.

TECHNICAL LITERATURE AND MANUALS

Technical manual – Bevplate trussed rafters, Bevplate Ltd, Rectory Farm Road, Sompting, Lancing, Sussex, BN15 0DP.
Trussed rafter constructions and specification guide – Gang-Nail Ltd, The Trading Estate, Farnham, Surrey, GU9 9PQ.
Trussed rafter construction information sheets – Hydro-Air International (UK) Ltd, Midland House, New Road, Halesowen, West Midlands, B63 3HY.
Technical manual – Trusswal Systems Ltd, 70 London Street, Reading, Berkshire, RG1 4SJ.
Technical manual – Twinaplate Ltd, Threemilestone, Truro, Cornwall, TR4 9LD.
Structural fixings – Bat Building and Engineering Products Ltd, Halesfield 9, Telford, Shropshire.
Moelven Toreboda Glulam – Moelven (UK) Ltd, 65 New Road, Netley Abbey, Southampton, Hants, SO3 5AD.
Technical roofing manual – Marley Ltd, Sevenoaks, Kent.
Redland Tiles technical manual – Redland Tiles Ltd, Reigate, Surrey, RH2 0SJ.
The development of roof space – The Velux Company Ltd, Telford Rd, Eastfield Industrial Estate, Glenrothes, Fife, KY7 4NX.

BOOKS

Timber Designer's Manual (to BS 5268 part 2) – J.A. Baird and E.C. Ozelton. The structural design of timber components and structures.

Roofing Ready Reckoner – Ralph Goss. Practical cutting tables for timber roof components.
Building in England – Sulzman. History of building development including roof structures.
Concise Encyclopaedia of Architecture – Martin and Briggs. Architectural styles.
Timber in Construction – TRADA. All forms of timber and sheet material usage in construction.
Site Carpentry – C.K. Austin. Practical site carpentry.

Index

Page numbers in italic type refer to diagrams.